6th International Workshop & Summer School on Plasma Physics 2014 (IWSSPP'14)

Journal of Physics: Conference Series Volume 768

Kiten, Bulgaria
30 June – 6 July 2014

Editors:

E. Benova
M. Atanasova

ISBN: 978-1-5108-3227-5
ISSN: 1742-6588

Printed from e-media with permission by:

Curran Associates, Inc.
57 Morehouse Lane
Red Hook, NY 12571

proceedings.com

Some format issues inherent in the e-media version may also appear in this print version.

Printed by Curran Associates, Inc. (2016)

For permission requests, please contact the Institute of Physics
at the address below.

Institute of Physics
Dirac House, Temple Back
Bristol BS1 6BE UK

Phone: 44 1 17 929 7481
Fax: 44 1 17 920 0979

techtracking@iop.org

Additional copies of this publication are available from:

Curran Associates, Inc.
57 Morehouse Lane
Red Hook, NY 12571 USA
Phone: 845-758-0400
Fax: 845-758-2633
Email: curran@proceedings.com
Web: www.proceedings.com

Table of contents

Volume 768

6th International Workshop & Summer School on Plasma Physics 2014 (IWSSPP'14)

30 June to 6 July 2014, Kiten, Bulgaria

Accepted papers received: 19 October 2016
Published online: 4 November 2016

Preface

Fusion Plasma and Materials

Plasma Modelling and Fundamentals

Plasma Sources, Diagnostics and Technology

Sixth International Workshop & Summer School on Plasma Physics 2014

Evgenia Benova *et al* 2016 *J. Phys.: Conf. Ser.* **VV**

The Sixth International Workshop & Summer School on Plasma Physics (IWSSPP'14) was organized by St. Kliment Ohridsky University of Sofia, with co-organizer PLASMER Foundation.

It was held in Kiten, Bulgaria, at the Black Sea Coast, from June 30 to July 6, 2014. The scientific programme covers the topics ***Fusion Plasma and Materials; Plasma Modeling and Fundamentals; Plasma Sources, Diagnostics and Technology***.

The Workshop ***Plasma for Sustainable Environment*** was co-organized together with the Institute of Plasmas and Nuclear Fusion, Lisbon, Portugal.

A special Workshop on ***Remote GOLEM operation*** was organized by the Institute of Plasma Physics, Prague, Czech Republic for the students and interested participants to work remotely with the Czech TOKAMAK GOLEM.

As with the previous issues of this scientific meeting, its aim was to stimulate the creation and support of a new generation of young scientists for further development of plasma physics fundamentals and applications, as well as to ensure an interdisciplinary exchange of views and initiate possible collaborations by bringing together scientists from various branches of plasma physics.

This volume of *Journal of Physics: Conference Series* includes 19 papers (invited lectures, contributed talks and posters) devoted to various branches of plasma physics, among them fusion plasma and materials, dc and microwave discharge modelling, transport phenomena in gas discharge plasmas, plasma diagnostics, cross sections and rate constants of elementary processes, material processing, plasma-chemistry and technology. Some of them have been presented by internationally known and recognized specialists in their fields; others are MSc or PhD students' first steps in science. In both cases, we believe they will raise readers' interest.

We would like to thank the members of both the International Advisory Committee and the Local Organizing Committee, the participants who sent their manuscripts and passed through the (sometimes heavy and troublesome) refereeing and editing procedure and our referees for their patience and considerable effort to improve the manuscripts. We would like to express our gratitude to the invited lecturers who were willing to pay the participation fee. In this way, in addition to the intellectual support they provided by means of their excellent lectures, they also supported the school financially.

We are obliged also to Bolkan Construction JSC for the sponsorship.

E. Benova and M. Atanasova

Peer review statement

All papers published in this volume of *Journal of Physics: Conference Series* have been peer reviewed through processes administered by the proceedings Editors. Reviews were conducted by expert referees to the professional and scientific standards expected of a proceedings journal published by IOP Publishing.

Journal of Physics: Conference Series **768** (2016) 012001 doi:10.1088/1742-6596/768/1/012001

Soft X-ray tomography in support of impurity control in tokamaks

J Mlynar[1], D Mazon[2], M Imrisek[1], V Loffelmann[1], P Malard[2], T Odstrcil[3], M Tomes[1], D Vezinet[3] and V Weinzettl[1]

[1] Institute of Plasma Physics AS CR, v.v.i., Za Slovankou 3, 182 00 Praha 8, Czech Republic
[2] IRFM, CEA Cadarache, 13108 Saint Paul lez Durance cedex, France
[3] Max Planck Institute for Plasma Physics, Boltzmannstr. 2, 85748 Garching, Germany

E-mail: mlynar@ipp.cas.cz

Abstract. This contribution reviews an important example of current developments in diagnostic systems and data analysis tools aimed at improved understanding and control of transport processes in magnetically confined high temperature plasmas. The choice of tungsten for the plasma facing components of ITER and probably also DEMO means that impurity control in fusion plasmas is now a crucial challenge. Soft X-ray (SXR) diagnostic systems serve as a key sensor for experimental studies of plasma impurity transport with a clear prospective of its control via actuators based mainly on plasma heating systems. The SXR diagnostic systems typically feature high temporal resolution but limited spatial resolution due to access restrictions. In order to reconstruct the spatial distribution of the SXR radiation from line integrated measurements, appropriate tomographic methods have been developed and validated, while novel numerical methods relevant for real-time control have been proposed. Furthermore, in order to identify the main contributors to the SXR plasma radiation, at least partial control over the spectral sensitivity range of the detectors would be beneficial, which motivates for developments of novel SXR diagnostic methods. Last, but not least, semiconductor photosensitive elements cannot survive in harsh conditions of future fusion reactors due to radiation damage, which calls for development of radiation hard SXR detectors. Present research in this field is exemplified on recent results from tokamaks COMPASS, TORE SUPRA and the Joint European Torus JET. Further planning is outlined.

1. Introduction

Fusion research has been pursued with the aim of achieving conditions required for useful production of energy by nuclear fusion in high temperature plasmas. These conditions are very severe even for the most reactive fusion fuel, i.e. the mix of hydrogen isotopes deuterium and tritium [1]. In the case of the magnetic confinement fusion (MCF) it is paramount to increase the energy confinement time, see e.g. [2]. This parameter is defined as follows:

$$\tau_E = \frac{W_P}{P_L} \ , \tag{1}$$

where W_P is thermal energy and P_L total power losses of plasma in a thermal equilibrium. It is presently the tokamak configuration that comes with the best performance in terms of energy confinement time. As a result, the ITER machine is a tokamak [3].

Most of the power losses in high-temperature hydrogen plasmas are due to the turbulent thermal convection, radiation losses prove to be much lower. However, the radiation losses may increase significantly due to the presence of heavy plasma impurities. First, intensity of bremsstrahlung (i.e. the free-free collisions between electrons and ions) rise with square of the ion charge number, and second, more importantly, impurities with high proton number Z will not get fully stripped, i.e. completely ionised to bare nuclei even in thermonuclear conditions, at temperatures around 150 million kelvin (approx. 15 keV). Therefore, with heavy impurities the recombination (the free-bound) and line (deexcitation, bound-bound) radiation sharply increases in intensity and exceeds the bremsstrahlung, see figure 1 (notice the logarithmic scales). This figure allows comparing intensities of hydrogen plasma bremsstrahlung with total radiation of tungsten impurity. The figure clearly demonstrates that for the plasma core the radiation of tungsten (in particular, its line radiation) presents the key contributor to total radiation losses even at concentration of tungsten in the order of 10^{-5} relative to the plasma density, see also section 4. As a consequence, heavy plasma impurities can cause considerable decrease of the energy confinement time (1) and prevent thermonuclear burn.

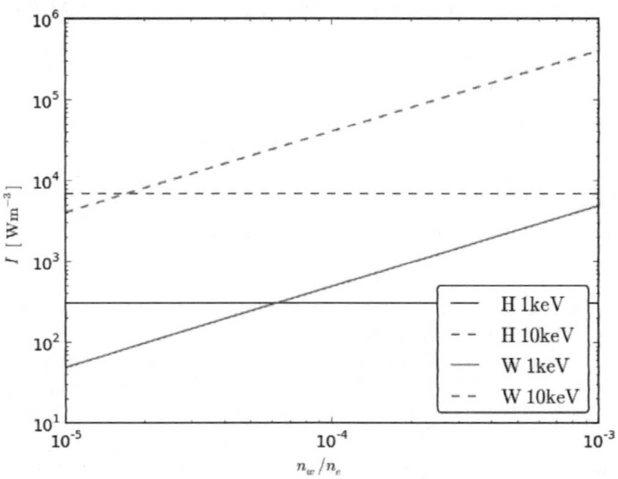

Figure 1. SXR radiation level from a high-temperature pure hydrogen plasma in comparison to total radiation of tungsten impurity as a function of the tungsten concentration at ion temperature 1 keV (full line) and 10 keV (dashed line)

The intensity and spectral properties of radiation losses from plasmas strongly depend on plasma temperature. In MCF experiments, plasma temperatures correspond to mean thermal energy of the plasma particles in the order of keV, which means that their radiation due to bremsstrahlung and atomic processes falls into the soft x-ray (SXR) frequency range. Tungsten with its proton number 74 deserves special attention here: Due to its high melting temperature and low sputtering it has been proposed as one of the materials for plasma facing components in ITER as well as in other tokamaks. The risk of the tungsten influx into the plasma core is therefore imminent and calls for dedicated studies. The fractional abundance of tungsten ionic charges for temperatures 1keV - 10keV is shown in figure 2, which is derived from the OPEN-ADAS database [4]. This figure demonstrates that even at the peak thermonuclear temperatures the tungsten charge is not higher than 57; therefore, its ions are a major contributor to the losses caused by radiation in the SXR region, and subsequently a serious concern to the energy confinement of fusion plasmas.

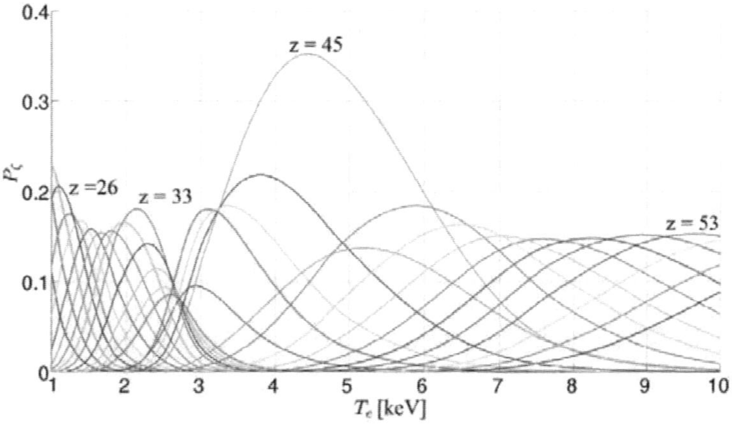

Figure 2. Fractional abundances of tungsten ions as a function of plasma temperature.

In the following section, principles of diagnostic methods for the SXR radiation of the MCF plasma core is briefly reviewed, including the novel Gas Electron Multiplier (GEM) detector [5]. In section 3, tomography reconstruction based on regularisation of the inversion task is presented and Singular Value Decomposition of the temporal evolution of the reconstructed emissivity exemplified on recent COMPASS results. In section 4, the present works aimed at use of SXR data for impurity control in real time are detailed, including SXR tomography reconstructions recently published by JET and by TORE SUPRA. In the conclusions, further planning in the framework of the co-ordinated fusion research is discussed.

2. SXR diagnostic methods

Spatial distribution of the SXR emission from MCF plasmas can be determined from a set of line-integrated measurements by detectors sensitive to SXR radiation. A typical diagnostic setup consists of a pinhole camera with a linear array of the SXR detectors. At present, semiconductor photosensitive elements are widely used for the SXR detection [6]. Recently, 2D pinhole cameras consisting of a matrix set of SXR detectors have been also proposed [7]. In order to limit the range of spectral sensitivity to the SXR region only, the pinhole is often combined with a spectral filter, usually in the form of a thin beryllium foil (tens to hundreds of micrometers thick), which is not transparent to ultraviolet and longer wavelength radiation. Notice that without this filter, the SXR intensity would be too low to contribute discernibly to the total observed intensity of plasma radiation. The filter limits the detected spectral region to photon energies higher than approx. 100 eV – 1 keV, while the high energy end is given by thickness of the sensitive region of the semiconductor photodiodes and can reach approximately 10 keV – 100 keV. The detectors can be absolutely calibrated in W/m^2 which corresponds to line integrated plasma power loss in the detector's spectral range. To improve sensitivity, stability and radiation resilience of the detectors, individual silicon barrier diodes, see e.g. [8], can be used instead of a the cost-effective set-up based on a photosensitive chip [6].

The semiconductor detectors - albeit radiation hardened - would not withstand conditions of the future fusion reactors including ITER due to the damage made by ionising radiation in the depleted area of the semiconductor. Therefore, a different principle of detection of the SXR radiation has been searched in order to design a reactor-relevant SXR diagnostic system, including e.g. diamond detectors or vacuum photodiodes. In the targeted spectral region, the Gas Electron Multiplier (GEM) represents a very promising candidate for a radiation resilient compact SXR detector with a good temporal and spatial resolution. GEM, which had been originally developed for particle physics research [5], [9], relies on conversion of the SXR photons to photoelectrons in a gas, followed by multiplication of the photoelectrons in a high electric potential in a grid made of a thin Kapton foil, see figure 3a. The photoelectrons are eventually collected at a pixelated anode, see figure 3b. The GEM detector works in a photon counting mode, with variable gain and, importantly, a tunable spectral response. The detector was successfully tested on TORE SUPRA [10], [11] where it is also foreseen as the principal diagnostics for the spatially resolved SXR studies in the future WEST configuration.

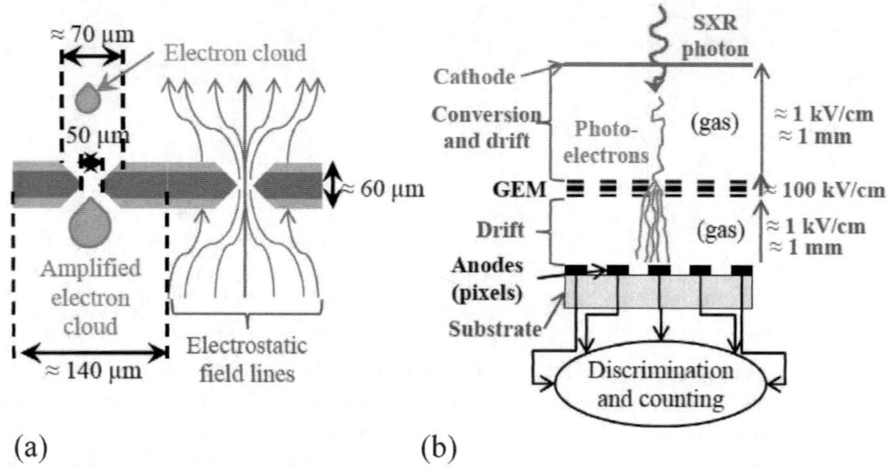

Figure 3. Principle of the Gas electron multiplier (GEM) detector of the SXR radiation: (a) detail of the Kapton foil with electron multiplication in the thin but intensive electric field (b) general scheme showing SXR photon, photoelectron, its multiplication in the grid made from the Kapton foil, and photon counting on the matrix anode. In practice, 'triple GEM' is used with three Kapton foils between the cathode and the anode. [10]

3. Tomography

Tomographic inversion is very often applied in the SXR data analyses in order to derive spatial distribution of SXR plasma emissivity from the measured line integrated data, i.e. from the plasma projections. The link between measured projections f_i and unknown distribution of the sources g_j is given by a set of linear equations

$$f_i = \sum_j^N T_{ij} g_j \quad , \tag{2}$$

where the contribution matrix T_{ij} specifies contribution of the j-th source of radiation to the i-th projection of plasma. It is obvious that this set of equations must be inverted to find the unknown g_j values, see figure 4. Inversion presents an ill-conditioned problem, so that existence and uniqueness of solution cannot be expected and minor errors in data can cause substantial errors in the resulting distribution of sources, called artefacts. In order to find a unique and dependable solution of an ill-conditioned problem, a suitable regularisation method has to be applied. A regularisation method usually relies in implementation of realistic constraints, in the case of plasma tomography e.g. smoothness of the distribution of the SXR radiation and its non-negativity.

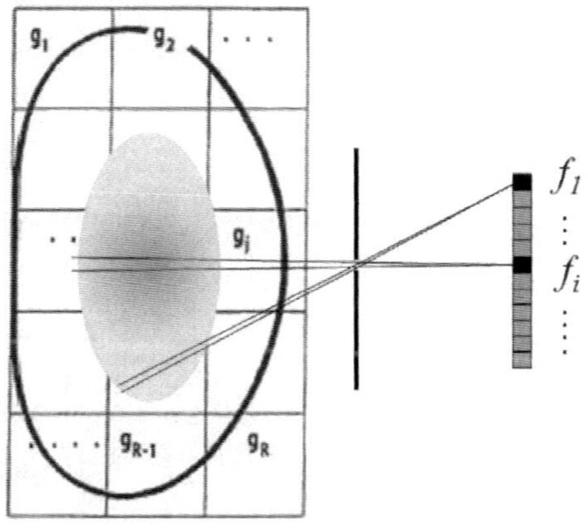

Figure 4. Principle of the pinhole camera and of division of the analysed plasma cross-section into pixels.

Plasma tomography in MCF is not only ill-conditioned, but also as a general rule underdetermined, as the spatial resolution of the SXR diagnostics is sparse. In particular, plasma projections can only be measured from a very limited set of angular positions due to constraints presented by magnetic coils, vacuum vessel and the in-vessel components. Several dedicated methods for MCF plasma tomography have been developed [12], [13]. In our experience, Minimum Fisher Regularisation (MFR) [14] algorithm based on Tikhonov regularisation constrained by minimisation of the Fisher information in the resulting image provides a robust and reliable solution [15].

In MFR, a reconstruction matrix M_{ji} is found that gives a unique smooth solution for the 2D plasma emissivity on a discrete rectangular mesh of R pixels (see figure 4) as follows:

$$g_j = \sum_{j}^{R} M_{ji} f_i \quad . \tag{3}$$

The matrix M_{ji} is found by Tikhonov regularisation which efficiently minimises $\Lambda_{MF} = \frac{1}{2}\chi^2 + \lambda\Omega$ where χ^2 is the goodness-of-fit parameter, λ is a regularisation (smoothing) parameter and Ω the objective functional [12]. In the MCF, the role of the objective functional is consigned to the Fisher information I_F defined by:

$$I_F = \int \frac{(\nabla g)^2}{g} dS \tag{4}$$

Due to the discrete character of eqs. (2) and (3) the I_F value has to be computed by advanced numerical methods. Besides, the Fisher information has to be implemented into the linear regularisation as an iterative process due to its nonlinear character [14]. Notice that in Tikhonov regularisation the smoothness of the resulting emissivity distribution can be controlled implicitly via expected data errorbars, i.e. through the goodness-of-fit χ^2.

As a significant novel physical constraint, anisotropic smoothness of the reconstruction with respect to the magnetic flux has been introduced in MFR via numerical differentiation matrices acting parallel and perpendicular to the magnetic flux, respectively. In this coordinate basis, preferential smoothness along the magnetic flux surfaces can be enforced, allowing for steeper gradients in the cross-field profile of plasma emissivity, for details see [15]. This new constraint reflects the fact that the plasma temperature and density gradients increase in the direction perpendicular to the field, and remain close to zero along magnetic field lines.

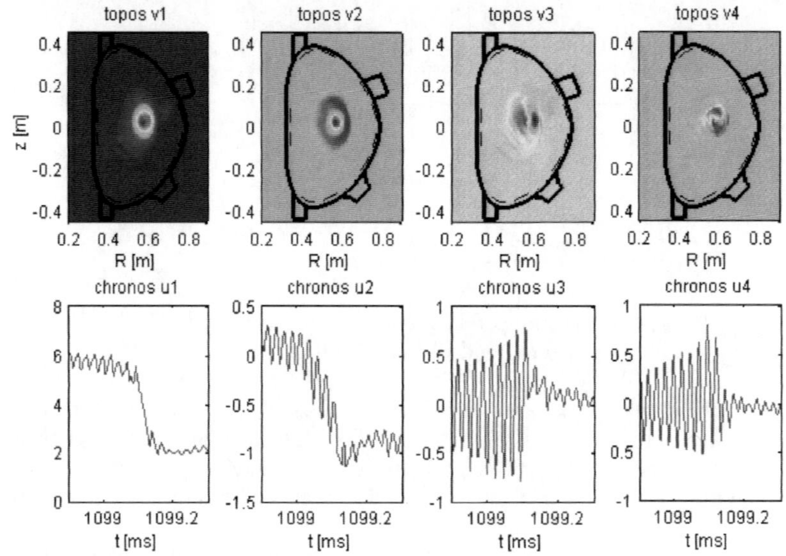

Figure 5. Singular value decomposition of a sawtooth event in the COMPASS discharge #6071. The thick black line in the topos image corresponds to the cross-section of the tokamak vacuum vessel.

Temporal evolution of plasma emissivity can be eventually decomposed into spatial and temporal eigenvectors - topos v_i and chronos u_i - by Singular Value Decomposition (SVD), see [14]:

$$g(r,t) = \sum_{i}^{\infty} s_i v_i(r) u_i(t) \quad . \tag{5}$$

The SVD helps to visualise the main processes and to filter out the noise. Performance of the SVD procedure is demonstrated in figure 5, which results from the MFR analyses of recent SXR measurements on the COMPASS tokamak. The SXR data have been collected from two pinhole detectors [6], each with a sufficient signal from the plasma core in about 12 lines of sight. In the figure, four eigenvectors showing SXR plasma emissivity (topos) with evolution of their amplitude (chronos) are presented, based on SVD from 114 timeframes of the tomographic reconstructions. The first and the second eigenvectors clearly display the effect of a sawtooth instability on the SXR emissivity – after the crash, the steep radiation peak in the plasma centre decreases in intensity, while the SXR intensity around the peak increases (notice that the second chronos has a negative amplitude). The third and the fourth eigenvectors visualise a rotating mode (notice the mutual phase shift in chronos), which increase in amplitude just before the sawtooth crash. Higher eigenvectors are not shown as they contain data noise.

Most of the present plasma tomography algorithms are computationally rather heavy and require input from processed data of other diagnostic systems, e.g. position and shape of magnetic field configuration. Although this is fully acceptable for the post-discharge data analyses, it disqualifies

Journal of Physics: Conference Series **768** (2016) 012001 doi:10.1088/1742-6596/768/1/012001

tomography from the list of prospective tools for the real-time control (RTC) of the tokamak discharge. Therefore, a new simplified version of real-time relevant MFR has been developed recently on the COMPASS tokamak [16]. The main novel idea relies in avoiding iterative processes and instead assuming smooth temporal evolution of data. The first applications of this new MFR version demonstrated good stability of the algorithm and a low level of artefacts. However, even in this version the computation load reaches present limits on the allowed delay time for the RTC. Therefore, further evolution of linear inversion methods is fostered on COMPASS together with developments of simplified rapid algorithms for non-inversion analyses of SXR data to directly calculate some basic parameters (e.g. plasma position) for the prospective RTC applications [17].

4. Impurity control: the way forward

In the SXR diagnostic applications focused on impurity studies, the information of key interest is the evolution of density distribution of the radiating species n_S. The SXR tomography reconstructs the evolution of plasma emissivity

$$g = n_e \sum_S n_S L_S \quad , \tag{6}$$

where the radiation cooling factor L_S – actually only its filtered part, corresponding to the detector spectral sensitivity range [18] – is a function of plasma temperature and impurity transport. Therefore, in order to derive impurity density from the SXR emissivity, plasma density n_e, temperature T_e and impurity transport properties of individual species must be accounted for. In the foreseeable future, this would not be feasible within the available delay times of the real-time control, in particular due to the complexity of the transport models. In other words, for a real-time relevant diagnostic system it is desirable to avoid SXR spectral intervals with radiation of ionic states for which transport plays a significant role.

Recent studies have shown that in SXR radiation of tokamak plasmas at high temperatures, transport plays little role for low Z impurities (which become fully stripped rapidly) and heavy Z impurities (which have very fast ionisation-recombination characteristic time scales [19]). The challenge thus remains with the intermediate Z impurities. A detailed study was published recently [18] where it is shown that the dependence of the filtered cooling factor on transport can be neglected provided that the spectral sensitivity of the SXR detector is limited to certain range of energies. However, it is far from obvious to find a suitable detection method for this task. Dedicated experiments with the GEM camera are currently planned to validate the possibility of controlling the GEM spectral sensitivity region just to the required range of energies. With this spectral filtering, the cooling factor would be dependent on local plasma temperature and density only. Consequently, a system for real-time control of impurity species could promptly derive evolution of the distribution of species n_S from real-time data on plasma density, plasma temperature and real-time SXR tomography. This is currently considered as a potentially feasible process.

At present, considerable efforts have been invested into analysing the special case of tungsten, a very heavy impurity with high probability of incident influx to the MCF plasma ('tungsten event') due to erosion of the plasma facing material. For heavy impurities, transport effects on the SXR radiation are negligible [18], [19]. However, it still proves rather demanding to derive tungsten density distribution from the SXR data due to systematic errors in diagnostic systems on one side (in particular, with respect to the SXR spectral sensitivity, or with mapping of the temperature profile) and in the atomic data on the other side. The contribution of line radiation to the overall intensity of radiation by tungsten can be computed with application of the photon emissivity coefficients provided e.g. by OPEN-ADAS [4] on the results of the ionization balance corresponding to the measured temperature profile. In tokamaks, the simplified calculation of this balance can be based on the collisional radiative steady state model, for details see [20], [21].

Journal of Physics: Conference Series **768** (2016) 012001 doi:10.1088/1742-6596/768/1/012001

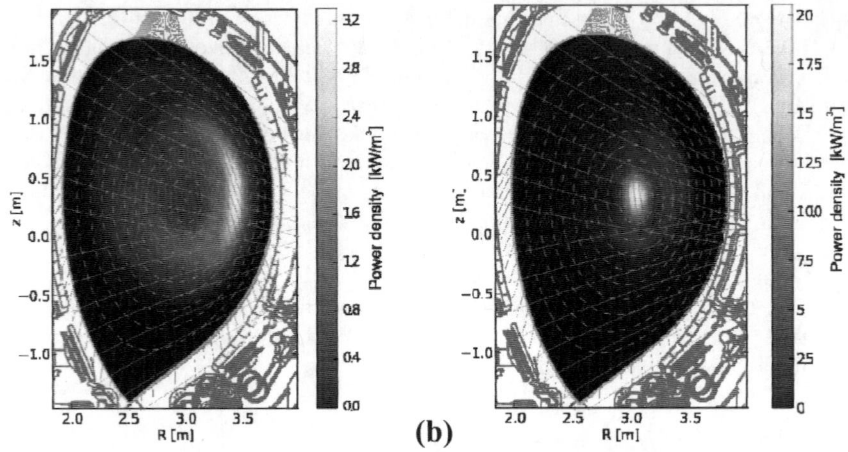

Figure 6. Tomographic reconstruction of the poloidal cross-section of the SXR emissivity during a tungsten event in JET discharge #82722 at two different times: (a) at 5.6 s, (b) at 7.5 s. [23]

SXR radiation due to tungsten impurity has been systematically studied in particular at JET where the ITER-like wall with tungsten divertor has been used since 2011. The diagnostic system for analyses of spatial distribution of the SXR radiation is challenging at JET for three reasons, see [22]: first, the pinhole cameras have different spectral sensitivities due to different thicknesses of Be foils, second, the cameras are located at different toroidal positions and third, the horizontal camera is tilted in toroidal direction. In spite of these difficulties, useful results have been achieved with the MFR algorithm [15], [22]. Recent results - which proved to be in good agreement with transport modelling, see [23], are exemplified in figure 6. In this figure, the radiation asymmetry at 5.6 s is due to centrifugal forces on tungsten in a toroidally rotating plasma. At a later time, at 7.5 s most of the tungsten impurity becomes accumulated in the plasma centre. Other JET works demonstrate that radiofrequency heating can be used to prevent this accumulation, e.g. [24].

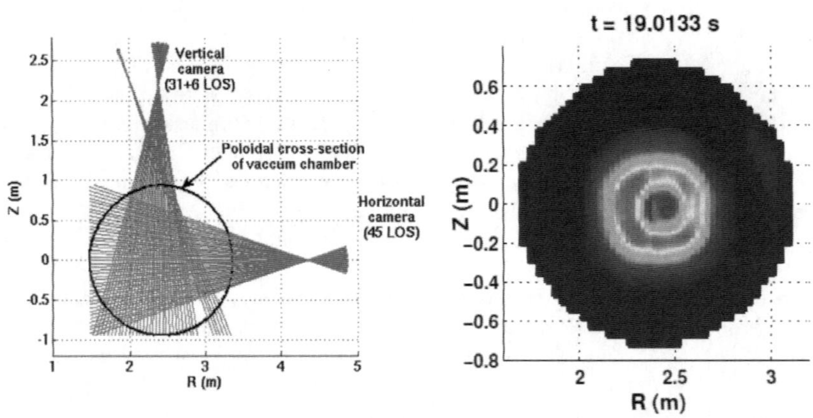

Figure 7. SXR tomography at TORE SUPRA. Left: experimental set-up with three pinhole cameras. Right: Cross-section of the SXR emissivity in the tungsten ablation experiment with LHCD in discharge #47757 shortly before tungsten radiation peaks in the plasma centre. Notice the asymmetry is opposite to the asymmetry due to centrifugal forces in figure 6. [27]

Even if a prompt and reliable sensor of behaviour of impurities existed, RTC would still require a process model and an efficient actuator to control plasma impurities – above all, to prevent influx of heavy impurities into the plasma core. It is beyond the framework of this contribution to discuss the challenge of developing actuators for the impurity control, see refs. [24] – [26]. Let us briefly summarise here that in general, plasma heating systems have proved to influence impurity transport in MCF plasmas. Ion cyclotron resonant heating (ICRH) has been routinely used to prevent impurity concentration in the plasma core, see e.g. [25]. In the basic interpretation, increased electron temperature exerts higher pressure on ions with multiple charge and results in a negative pinch [26], however, quantitative understanding is still incomplete so that the present use of ICRH in the impurity control is largely empirical. A clear effect of Lower Hybrid Resonant Heating on impurity radiation was also observed, see figure 7. Understanding this result proves even more challenging [27]. To conclude, an in-depth investigation of the prospective actuators and further development of the process model is required before the impurity control can make a standard part of the RTC system. Dedicated experimental campaigns will be pursued at major facilities, including ASDEX Upgrade and JET in order to benchmark simulations and increase our understanding of the impurity transport.

5. Conclusions

SXR diagnostic with spatial and temporal resolution proves to be information rich, however analysis and interpretation of the information is not straightforward and can be computationally heavy for multiple reasons. Among others, tomographic inversion to retrieve distribution of radiation sources from plasma projection is an ill-conditioned task. Next, SXR intensity can be dominated by impurity radiation, which in turn can be dependent on plasma transport properties.

On the other hand, the SXR diagnostic systems have unique capability of diagnosing heavy impurities in the MCF plasmas with valuable (albeit limited) spatial and high temporal resolution. Therefore, SXR diagnostics present a candidate sensor for future real-time impurity control systems. With respect to this prospective, new simplified methods of SXR data analyses are under development at COMPASS tokamak, where engineering feasibility studies of SXR implementation into RTC are foreseen. In parallel, novel diagnostic methods with controllable spectral range and a high radiation resistivity are under development in CEA France in a framework of a broad international collaboration. While the TORE SUPRA tokamak is being refurbished to the WEST configuration with a tungsten divertor, the GEM detector shall be tested in Germany at ASDEX-Upgrade, a tokamak with the plasma facing wall made of tungsten. Further development of the atomic database will be also significant in this respect. It is foreseen that results from (i) real-time relevant data analyses, (ii) novel diagnostic system based on GEM and (iii) tungsten plasma facing components shall be merged in the future operations of the WEST configuration of TORE SUPRA. The Joint European Torus JET presents another essential facility for the tungsten impurity studies preceding the ITER operation, due to the ITER-like wall, high energy confinement and tritium capability of JET.

Acknowledgements
This work was supported by grant of the Czech Science Foundation GA P205/10/2055, by the MSMT project LG14002 and by the MSMT project LM2011021.

References:

[1] Lawson J D 1957 Some Criteria for a Power Producing Thermonuclear Reactor *Proc. Phys. Soc.* **B70** p 6
[2] Braams C M, Stott P E 2002 *Nuclear Fusion, Half a Century of Magnetic Confinement Fusion Research* (IoP Publishing Ltd., ISBN 0750307056)
[3] ITER: the world's largest Tokamak, http://www.iter.org/mach cit. June 9, 2014
[4] http://open.adas.ac.uk/ cit. June 24, 2014
[5] Sauli F 1997 GEM: A new concept for electron amplification in gas detectors *Nucl. Instrum. Methods Phys. Res. A* **386** pp 531-534

[6] Weinzettl V et al. 2010 Design of multi-range tomographic system for transport studies in tokamak plasmas *Nucl. Instrum. Methods Phys. Res. A* **623** p 806

[7] Murari A et al. 2012 Latest developments in image processing for the next generation of devices with a view on DEMO *Fusion Eng. Des.* **87** p 2116

[8] Gill R D, Alper B and Edwards A W 1997 First Results in D-T from the Radiation Hardened Soft X-ray Cameras *JET Joint Undertaking report JET-R(97)11*

[9] Romano A et al. 2010 Characterization of a 2D soft x-ray tomography camera with discrimination in energy bands, *Rev. Sci. Instrum.* **81** p 10E523

[10] Vezinet D. 2013 *Soft X -ray Measurement and Analysis on Tokamaks in view of real-time control* (PhD thesis, CEA-IRFM France)

[11] Mazon, D et al. 2013 Soft X-ray imaging techniques on Tore Supra: Present status and possible future developments *Nucl. Instrum. Methods Phys. Res.A* **720** p 78

[12] Ingesson, L C, Alper B, Peterson B J, Vallet J-C 2008 Tomography Diagnostics: Bolometry and Soft X-Ray Detection *Fusion Sci. Technol.* **53** p 528

[13] Mlynar J, Weinzettl V, Bonheure G, Murari A and EFDA-JET contributors 2010 Inversion techniques in the Soft X-Ray tomography of fusion plasmas: towards real-time applications *Fusion Sci. Technol.* **58** p 733

[14] Anton M et al. 1996 X-Ray Tomography on the TCV Tokamak *Plasma Phys. Control. Fusion* **38** p 1849

[15] Odstrcil M, Mlynar J, Odstrcil T, Alper B, Murari A and JET-EFDA Contributors 2012 Modern numerical methods for plasma tomography optimisation *Nucl. Instrum. Methods Phys. Res. A* **686** p 156

[16] Löffelmann V 2012 *Soft X-ray tomography for the real time control of tokamak experiments* (BSc. thesis in Czech, FNSPE CTU Prague)

[17] Imrisek M et al. 2014 Use of soft x-ray diagnostic on the COMPASS tokamak for investigations of sawteeth crash neigborhood and of plasma position using fast inversion methods, 20th HTPD conference, *Rev. Sci. Instrum.*, submitted

[18] Vezinet D et al, Impurity density derivation from bandpass soft x-ray tomography: applicability, perspectives and limitations, Nucl. Fusion 54 (2014) 083011

[19] Puetterich T, Neu R, Dux R, Whiteford A D and O'Mullane M G 2008 Modelling of measured tungsten spectra from ASDEX Upgrade and predictions for ITER *Plasma Phys. Control. Fusion* **50** p 085016

[20] Puetterich T 2010 Calculation and experimental test of the cooling factor of tungsten *Nucl. Fusion* **50** p 025012

[21] Mlynar J, Tomes M et al. 2014 Soft X-ray tomographic reconstruction of JET ILW plasmas with tungsten impurity and different spectral response of detectors (28th Symposium on Fusion Technology, San Sebastian, Spain, accepted)

[22] Mlynar J et al. 2011 2D tomography of SXR data from toroidally separated cameras for studies of impurity injection and fast instabilities on JET (38th EPS Conf. Plasma Phys., Strasbourg, France) ECA **35G** P4.052

[23] Angioni C et al 2014 Tungsten transport in JET H-mode plasmas in hybrid scenario, experimental observations and modeling *Nucl. Fusion* (accepted)

[24] Jacquet P et al. 2014 Ion cyclotron resonance frequency heating in JET during initial operations with the ITER-like wall *Phys. Plasmas* **21** p 061510

[25] Valisa M. et al. 2011 Metal impurity transport control in JET H-mode plasmas with central ion cyclotron radiofrequency power injection *Nucl. Fusion* **51** p 033002

[26] Angioni C and Peeters A.G. 2006 Direction of Impurity Pinch and Auxiliary Heating in Tokamak Plasmas *Phys. Rev. Lett.* **96** p 095003

[27] Vezinet D. et al. 2014 LHCD-induced Soft X-Ray poloidal asymmetry during tungsten trace injection: investigating radiative processes (41st EPS Conf. Plasma Phys., Berlin, Germany) *ECA* **38** P5.056

Journal of Physics: Conference Series **768** (2016) 012002 doi:10.1088/1742-6596/768/1/012002

Remote operation of the GOLEM tokamak with hydrogen and helium plasmas

V Svoboda[1], A Dvornova[2], R Dejarnac[3], M Prochazka[4], S Zaprianov[5], R Akhmethanov[7], M Bogdanova[8], M Dimitrova[3,6], Zh Dimitrov[5], O Grover[1], L Hlavata[4], K Ivanov[5], K Kruglov[7], P Marinova[6], P Masherov[7], A Mogulkin[7], J Mlynar[1,3], J Stockel[1,3,4], A Volynets[8]

[1]Faculty of Nuclear Physics and Physical Engineering CTU, Prague, Czech Republic
[2]Moscow Engineering Physics Institute (MePhi), Moscow, Russian Federation
[3]Institute of Plasma Physics, AS CR, Za Slovankou 3, 182 00 Prague, Czech Republic
[4]Faculty of Chemistry, Brno University of Technology, Brno, Czech Republic
[5]St. Kliment Ohridski University of Sofia, Bulgaria
[6]Acad. E. Djakov Institute of Electronics, Bulgarian Academy of Sciences, Sofia, Bulgaria
[7]Research Institute of Applied Mechanics and Electrodynamics of Moscow Aviation Institute (National Research University), Moscow, Russian Federation
[8]Faculty of Physics, M.V.Lomonosov Moscow State University, Moscow, Russian Federation

E-mail: stockel@ipp.cas.cz

Abstract. The GOLEM tokamak was operated remotely via Internet connection during the 6th International Workshop and Summer School on Plasma Physics. Performances of hydrogen and helium discharges are compared in this paper. It is found, at similar vacuum conditions, that helium discharges are shorter but the breakdown of the working gas can be quite easily achieved at almost the same loop voltage. The plasma current in helium discharges is slightly lower than in the case of hydrogen. Turbulent fluctuations of the floating potential measured by means of an array of Langmuir probes reveal a noticeably different character in the two discharges.

1. Introduction

Remote operation of the GOLEM tokamak, which is located at the Czech Technical University in Prague, was performed on-line by twelve participants of the 6th International Workshop and Summer School on Plasma Physics in Kiten, Bulgaria. The plasma discharges were produced in hydrogen and helium as working gases and a comparison of performances of both types of discharges is presented in this paper. The GOLEM tokamak together with its state-of-the-art remote handling system are briefly described in section 2. Experimental results, including the optimization of the helium discharges, and the comparison of floating potential fluctuations measured by Langmuir probes (LPs), are presented in section 3.

[4] To whom any correspondence should be addressed.

Journal of Physics: Conference Series **768** (2016) 012002 doi:10.1088/1742-6596/768/1/012002

2. The GOLEM tokamak

2.1 Description and characteristics

The GOLEM tokamak is operational at the Faculty of Nuclear Physics and Physical Engineering (FNPPE), Czech Technical University in Prague [1]. GOLEM is a small tokamak which was constructed at the end of 1950's at the Kurchatov Institute, Moscow as TM-1. The tokamak was moved to the Institute of Plasma Physics in Prague in 1977 and re-named CASTOR [2]. After 30 years of operation, the tokamak was given to the FNPPE for education of students and renamed GOLEM.

The GOLEM tokamak has a circular cross section. The major/minor radii of the tokamak vessel are R_0 = 0.4 m, b= 0.1 m. The stainless steel vessel is equipped with a poloidal limiter (made of Molybdenum) of radius a = 0.085 m. The power supplies of individual windings are based on several condenser banks. Here, the condenser banks to supply the toroidal field coils and primary winding of the air core transformer are exploited.

The tokamak is equipped by a set of simple diagnostics, which measure the loop voltage, plasma current, toroidal magnetic field, and visible emission. GOLEM is also equipped with Mirnov coils, a visible spectrometer, an array of bolometers, a fast camera for time resolved pictures, etc. In the series of experiments described here, a radial array of 12 Langmuir probes is used.

Engineering and plasma parameters, which can be achieved on GOLEM are quite modest. The tokamak operates at maximum toroidal magnetic field of up to 0.5 T. The central electron temperature is less than 100 eV, the maximum line average density ~ 10^{19} m^{-3}, the maximum pulse length is around 18 ms.

2.2 Remote handling of the GOLEM tokamak

A unique capability of the GOLEM tokamak is that it can be operated remotely via Internet [3]. Once agreed with the chief operator, the users connect to the web page displaying the remote control room of GOLEM, which is displayed as print screen in Figure 1.

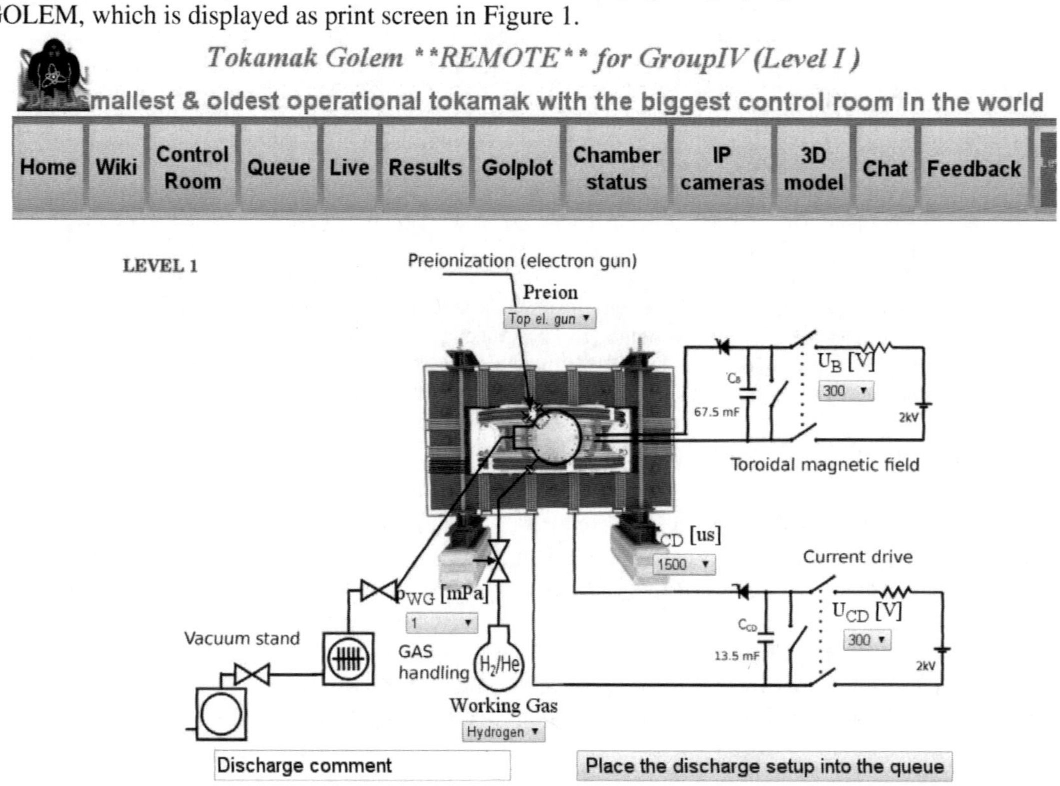

Figure 1. Virtual control room of the Golem tokamak used for remote operation.

Just six "buttons" shown in figure 1 are used to operate the tokamak. Participants select charging voltage of the condenser banks for powering the toroidal field coils (U_B) and the primary winding of the transformer (U_{CD}). Then, the time delay between trigger pulses of U_B and U_{CD} is also pre-selected (t_{CD}). Furthermore, the working gas (Hydrogen or Helium) and its filling pressure (p_{WG}) are chosen. One can also select the type of pre-ionization (microwave or electron gun). The selected discharge is commented and placed into the queue. Once the discharge is executed, the experimental results in form of temporal evolutions of basic plasma parameters, as well as resulting data files are available, when the option "Results" is selected on the yellow banner of the screen. Other knobs seen on the yellow banner in figure 1: Queue – position of the discharge to be executed in the queue, Live – views by web cameras of the torus hall and through a glass window into the tokamak vessel.

3. Experimental results

The remote operation of GOLEM from Kiten is focused on the comparison of discharge performances in Hydrogen and Helium plasmas. In the past, the majority of GOLEM discharges were performed with Hydrogen as the working gas. However, some features of plasma performance in tokamaks are related to so-called isotope effects. This is important not only with respect to plasma physics but it has some practical consequences for ITER operation, since a campaign in Helium is planned in ITER after the initial phase in hydrogen. Some features like the threshold power for L-H transition in Helium plasmas should be predicted with sufficient precision well in advance. Therefore, the first task of the remote operation was devoted to optimization of He discharges. The second task is to compare the edge plasma turbulence in hydrogen and helium plasmas.

3.1 Optimization of helium discharges

First of all, an optimum set of input parameters had to be determined to get stable Helium discharges with the lowest loop voltage at the breakdown and the highest plasma current. It was found that a key parameter to get stable He discharge is to select a sufficiently high filling pressure of the working gas, as documented in figure 2.

Figure 2 displays temporal evolutions of the loop voltage for three discharges differing in the value of the filling pressure, with remaining input parameters being the same (U_B = 800 V, U_{CD} = 450 V, t_{CD} = 2 ms). We clearly see that a sufficiently high pressure of He is required to achieve a stable discharge without MHD instabilities. It has to be noted that the loop voltage is at acceptable level in all three cases with a break down less than 10 V, resulting in long discharges.

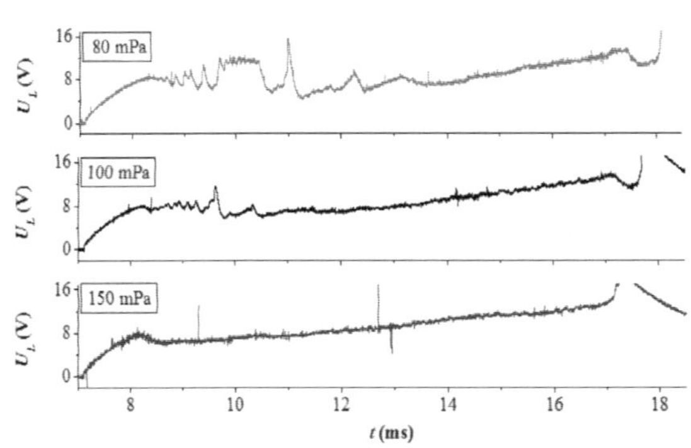

Figure 2. Evolution of the loop voltage of helium discharges at different pressures.

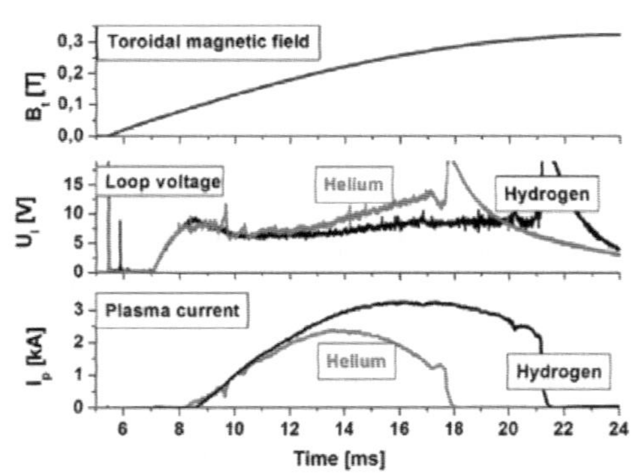

Figure 3. Comparison of the loop voltage and plasma current in He (#16339) and H (#16319) discharges at the filling pressure p =100 mPa.

The next task is to find discharges in Hydrogen with similar performance. Figure 3 compares temporal evolutions of the loop voltage and plasma current for two discharges in Helium and Hydrogen.

It has to be noted that operation in He or Hydrogen is preceded by glow discharge cleaning in Helium or Hydrogen, respectively. To find two identical discharges in both working gases was challenging and the best-achieved result is for shot #16319 and #16339 for time t < 13 ms.

3.2 Fluctuation measurements

Floating potential is measured using a radial array of Langmuir probes (the so-called rake probe) [4], which is shown in figure 4.

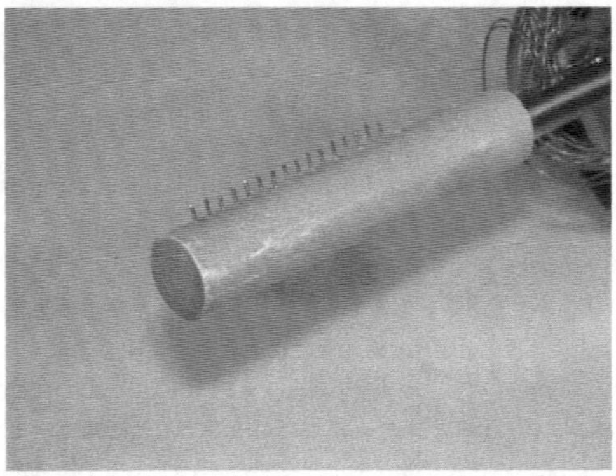

Figure 4. Picture of the rake probe. The insulating probe head is made of Boron Nitride.

The rake probe consists of 16 molybdenum tips with diameter of 0.7 mm and length of 2 mm. However, only 12 tips are used in this experiment because of limitation of available data acquisition channels. The rake probe is inserted into the plasma from the bottom of the vessel. The first Langmuir probe (LP1) is the deepest located at r = 70 mm from the center of the tokamak vessel. The distance between the individual probes is 2.5 mm. Consequently, the probe LP7 is located at radius r = 85 mm, which corresponds to the radius of the GOLEM limiter, so probes LP8 – LP12 are in the limiter shadow with open magnetic field lines. The probe LP9 appears to be out of operation during described experiments. Probe signals are digitized at 1 MHz sampling rate and stored in the GOLEM database.

About 53 discharges (#16293 - #16346) were executed during the remote session from Kiten. To compare properties of turbulent fluctuations in Helium and Hydrogen plasmas, we selected two discharges, which are characterized by a similar evolution of the loop voltage. Figure 5 compares discharges in Hydrogen (#16312) and in Helium (#16346). The temporal evolution and the maximum toroidal magnetic field are identical $B_T = 0.33$ T ($U_B = 800$ V) for both discharges. The capacitor bank for primary winding of the transformer is charged to U_{CD} =300 V (H) and U_{CD} = 200 V (He), respectively. It has to be noted that these values are below charging voltages usually used on GOLEM operation. The filling pressure of Helium is roughly twice that (79 mPa) of the H discharge (35 mPa).

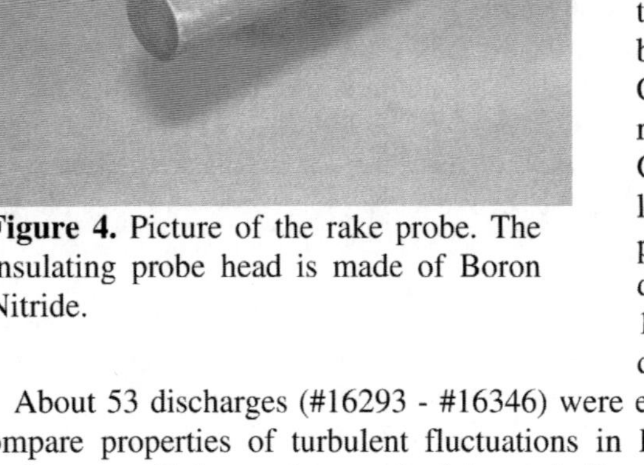

Figure 5. Comparison of the loop voltage and plasma current in and H (#16312) and He (#16346) discharges.

It is seen in the figure that the loop voltage required for plasma breakdown is quite low in both discharges, 5.8 V for hydrogen plasma and even lower for helium discharge, 4.7 V. It has to be noted that the breakdown occurs at a quite low toroidal magnetic field, $B_T = 0.052$ T. However, the maximum value of the plasma current differs significantly in these discharges. In hydrogen plasma, the maximum plasma current is 2.5 kA, while only 0.97 kA is achieved in Helium plasma. In both cases, the discharges are stable, without any evident MHD instabilities. The discharge in Helium is shorter by about 2.5 ms than the Hydrogen one. Such shortening of He discharges is observed for all discharges in this remote operation campaign.

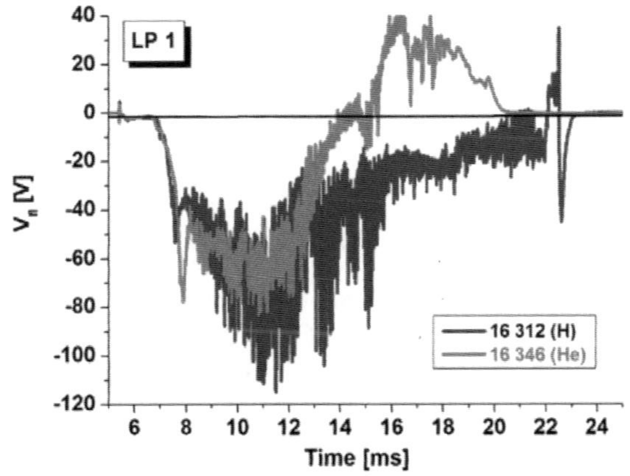

Figure 6. Comparison of the floating potential recorded by the probe LP1 in Hydrogen and He plasmas.

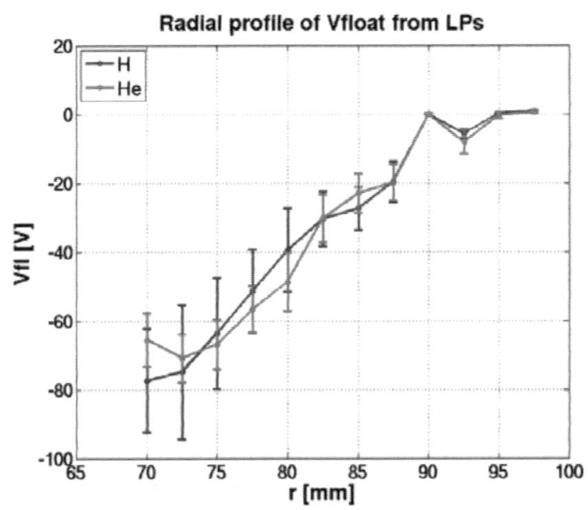

Figure 7. Radial profiles of the floating potential in Hydrogen (blue) and Helium (red) discharges. Probe LP9 is out of operation.

The signals of the floating potential of the probe LP1 are compared in figure 6. It is seen that both probe signals are time dependent. The most probable explanation of such a variation of the V_{fl} mean values is a vertical (and also radial) movement of the plasma column during the discharge, GOLEM being not equipped with any feedback control for position. At the beginning of the He discharge, the floating potential is negative, which is typical for the probe located deep in the confined region of the plasma column [5], while the V_{fl} is positive for $t > 16$ ms. Therefore, we speculate that the plasma column moves from bottom to top during the discharge. Unfortunately, this speculation cannot be confirmed by magnetic diagnostics [6] or by fast tomography [7], because these diagnostics were out of operation during this experimental campaign. Therefore, to analyze turbulent fluctuations and to compare their H/He properties, we focus on a short time interval during the discharge where V_{fl} is relatively in steady state. We select a short time window of duration of $\Delta t = 1$ ms at time $t = 10.5$ ms when the floating potential is minimum.

The radial profiles of the floating potential in Hydrogen and Helium discharges are compared in figure 7. The data are time averaged over Δt and the error bars correspond to the standard deviation around the mean value. In spite of different basic discharge parameters the radial profiles are almost identical. The slope of the profile $-dV_{fl}/dr$ is proportional to the radial electric field (if we neglect the unknown gradient of the electron temperature) and is ~1.25 kV/m in both cases. Such value of the radial electric field causes a significant ExB velocity in the poloidal direction, which is around v_{pol}~1.25/0.16~7.8 km/s (with Bt = 0.16 at t = 10.5 ms). We note opposite gradients of V_{fl} between the probes LP1 and LP2 for the two discharges.

Figure 8 compares the fluctuation component of the floating potential as measured by LP1 in Hydrogen and Helium discharges. It is evident that the level of fluctuations is noticeably different for the entire duration of the discharges, significantly smaller in the case of He plasma, for the deepest probe.

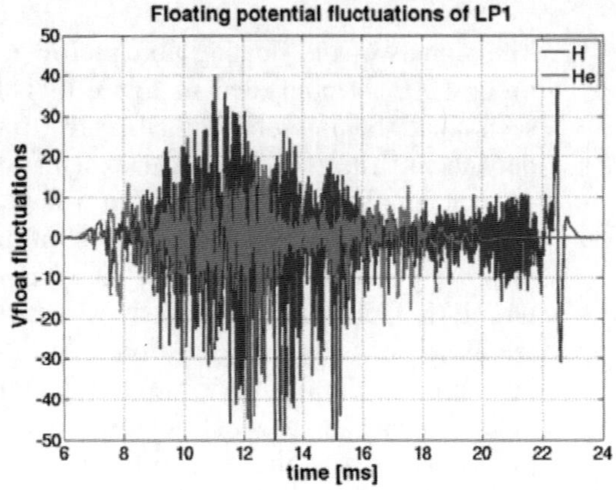

Figure 8. Fluctuations of the floating potential in Hydrogen (#16312-blue) and Helium (#16346-red) discharges as recorded by the probe LP1.

A detailed comparison of the V_{fl} fluctuations properties for all LPs is presented in figure 9 with the Probability Distribution Function (PDF) as measured by probes LP1 – LP8.

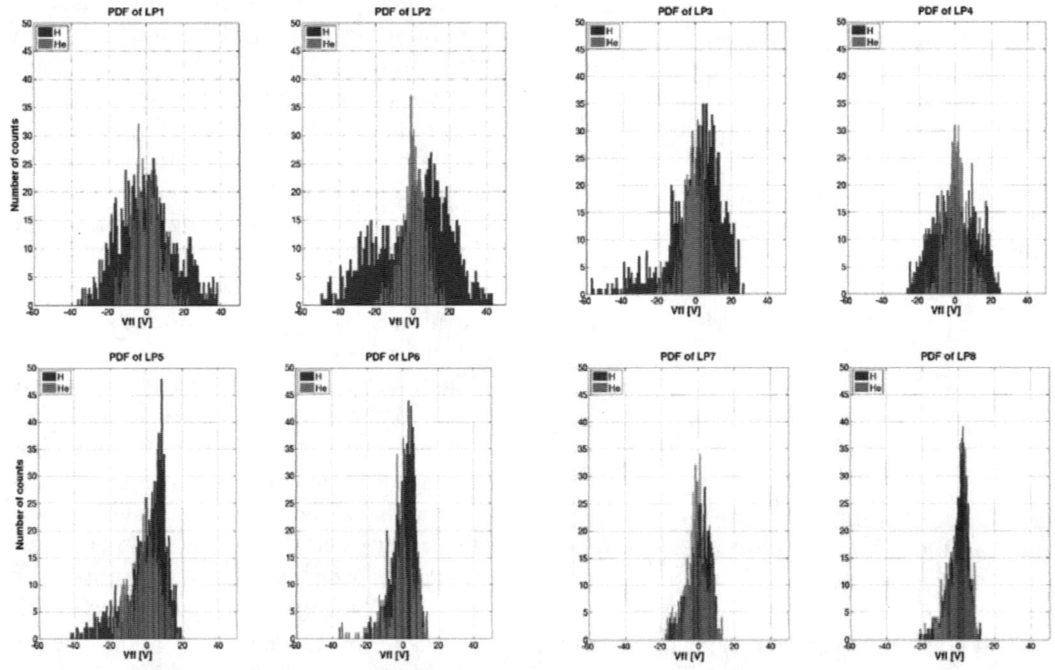

Figure 9. PDF of V_{fl} fluctuations in H (#16312) and He (#16346) discharges for probes LP1 –LP8.

It is evident that the shape of PDF depends on the probe position inside the plasma column. The PDFs in Hydrogen are broader than in Helium, except the probes LP7 and LP8, which are located in the limiter shadow. The Probability Distribution Functions look mostly Gaussian in Helium plasma, while negative tails in the floating potential are evident in the Hydrogen discharge.

Figure 10 compares the power spectra in H and He for the same probes.

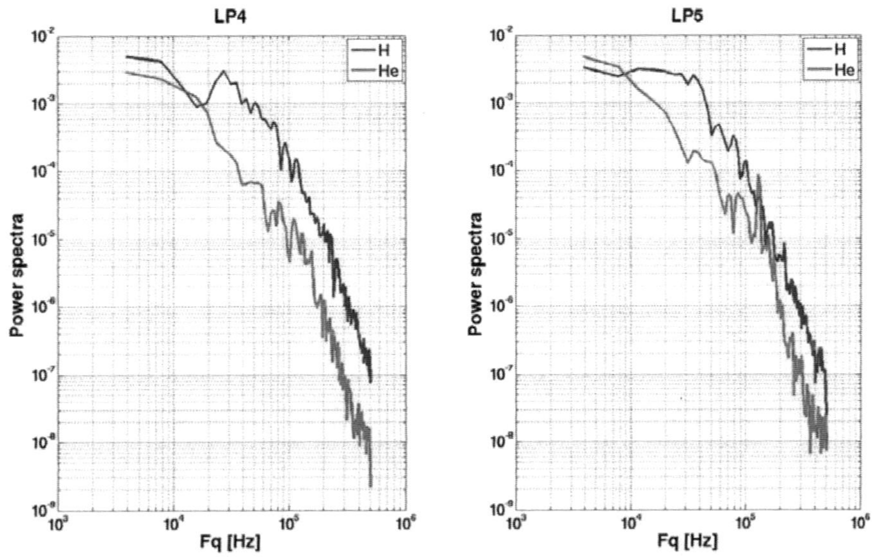

Figure 10. Power spectra of V_{fl} fluctuations in H and He discharges for probes LP4 (left) and LP5 (right).

A characteristic peak of fluctuation power is seen in hydrogen plasma at f = 28 kHz, which is missing in Helium. A high frequency peak in Helium plasma localized around 120 kHz is present.

Figure 11 displays an example of the cross correlation between two probes, radially spaced by d = 2.5 mm.

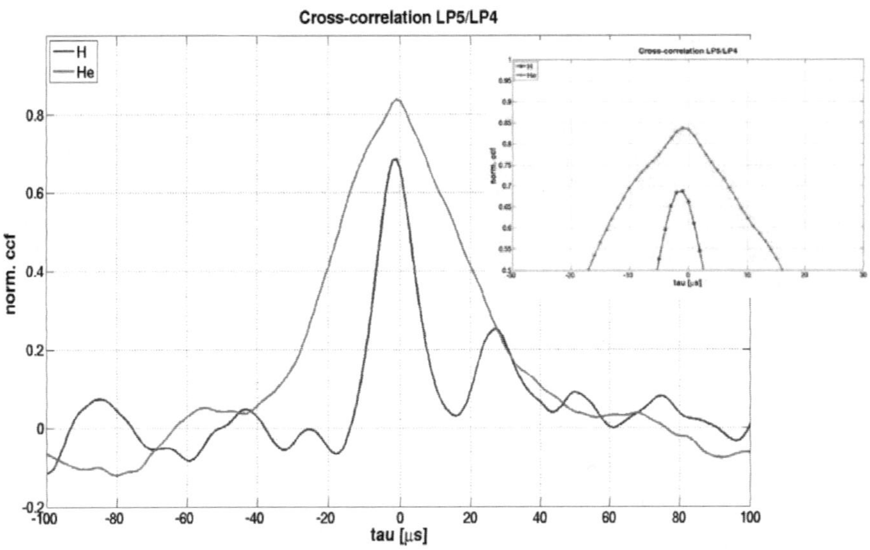

Figure 11. Cross correlation between probes 4 and 5 in H and He discharges. Zoom around the maximum is inserted.

The cross correlation between probes LP4 and LP5 is significant in H and He discharges, being 70 - 80%. The negative time lag, τ = -1 μs at the maximum of the cross correlation function, evident from the insert in figure 8, would imply a radial propagation of turbulent structures from LP4 to LP5. The velocity of this turbulent structure, or blob [8], can be simply estimated as v_radial = d/τ = 2.5 km/s.

4. Conclusions

We demonstrate here that the GOLEM tokamak is a unique facility, which is effectively used for motivating students for fusion research and their practical training anywhere in the world in a simple, remote way. Furthermore, interesting experimental results can be achieved. The remote operation of GOLEM from the Kiten workshop was focused on comparative studies of plasma performance by using Hydrogen and Helium as working gases. We clearly show that Helium plasma is easily generated with plasma parameters comparable with Hydrogen plasma, which is not standard in other tokamaks. Therefore, this feature allows the study of mass composition effects, which might be important for larger tokamak facilities. We focus on the comparison of turbulent fluctuations properties of the floating potential in H/He plasmas. The floating potential is measured by a radial array of 12 Langmuir probes covering the limiter shadow as well as part of the confined plasma. We demonstrate that in He plasma the level of V_{fl} fluctuations is noticeably lower, that the Probability Distribution Function is closer to a Gaussian, and that the frequency spectra differ from those measured in Hydrogen plasma. However, one question still remains open – the role of plasma density on fluctuation properties in H/He plasmas. We cannot exclude that the plasma density is similar in these two discharges under discussion, because the plasma density was not measured during this experimental campaign. Nevertheless, the filling pressure for the analyzed discharges was almost the same and recent interferometric measurements on GOLEM show quite similar values of the line average density under this condition.

It is evident that additional experiments have to be performed and more sophisticated analyses to be exploited to get better insight of these mass composition effects on properties of the plasma turbulence.

Acknowledgements

This work was partially supported by the project RVO: 68407700 of the Czech Technical University in Prague. Additional support has been granted from IAEA technical contract CRP F1.30.14 on "Utilization of the Network of Small Magnetic Confinement Fusion Devices for Mainstream Fusion Research". The opinions expressed by authors do not necessarily represent the positions of the European Commission neither IAEA.

References

[1] Svoboda V, Pokol G, Réfy D I, Stöckel J, Vondrášek G, Former tokamak CASTOR becomes remotely controllable GOLEM at the Czech Technical University in Prague, *37th EPS Conference on Plasma Physics*, Dublin, June 2010

[2] Stockel J, Badalec J, Duran I, Hron M, Horacek J, Jakubka K, Kryska L, Petrzilka J, Zacek F, Heller M V P, Brazilio Z A, Caldas I L 1999 *Plasma Phys. Control. Fusion* **41** A577-A585 Suppl. 3A

[3] Svoboda V, Huang B, Mlynar J, Pokol G I, Stöckel J, Vondrasek G 2011 *Fusion Engin. Design* **86** 1310–1314

[4] Stöckel J, Adamek J, Balan P, Bilyk O, Brotankova J, Dejarnac R, Devynck P, Duran I, Gunn J P, Hron M, Horacek J, Ionita C, Kocan M, Martines E, Panek R, Peleman P, Schrittwieser R, Oost G Van, Zacek F 2006 *J. Phys.: Conf. Series* **63** 012001

[5] Brotankova J, Stockel J, Horacek J, Seidl J 2009 *Plasma Physics Reports* **35**, No. 11, 980–986.

[6] Ficker O, Grover O, Kocman J, Krbec J, Loffeman V, Markovic T, Matusu M, Stockel J, Svoboda V, Veverka J, Vondrasek G, Tokamak GOLEM for fusion education chapter 5, *EPS 41st Conference on Plasma Physics*, Berlin, June 2014, P4.141

[7] Odstrcil T, Odstrcil M, Grower O, Svoboda V, Duran I, Mlynar *J* 2012 *Rev. Sci. Instrum.* **83**, 10E505 ; http://dx.doi.org/10.1063/1.4731003

[8] Krasheninnikov S, Ryutov D, Guanghuiet Y U 2004 *J. Plasma Fusion Res. Series* **6** 139-143

Journal of Physics: Conference Series **768** (2016) 012003 doi:10.1088/1742-6596/768/1/012003

Determination of the plasma potential and the EEDF by Langmuir probes in the divertor region of COMPASS tokamak

P Ivanova[1], M Dimitrova[1,2], E Vasileva[1], Tsv K Popov[3], R Dejarnac[2], J Stöckel[2], M Imríšek[2], P Hacek[2] and R Panek[2]

[1]Acad. E. Djakov Institute of Electronics, Bulgarian Academy of Sciences
72 Tsarigradsko Chaussee, 1784 Sofia, Bulgaria
[2]Institute of Plasma Physics, Academy of Sciences of the Czech Republic v.v.i.
3 Za Slovankou, 182 00 Prague 8, Czech Republic
[3]Faculty of Physics, St. Kl. Ohridski University of Sofia
5 J. Bourchier Blvd., 1164 Sofia, Bulgaria

E-mail: poly.ivanova@abv.bg

Abstract. This paper reports experimental data obtained on the COMPASS tokamak by an array of Langmuir probes embedded in the divertor tiles. The measured current-voltage probe characteristics were processed by the recently published first-derivative probe technique for precise determination of the plasma potential and the electron energy distribution function (EEDF). The measurements were performed during L-mode hydrogen and deuterium plasma with a toroidal magnetic field B_T = 1.15 T, plasma current I_p = 180 kA and average electron density $n_e = 8\times10^{19}\,m^{-3}$. The spatial profile of the electron temperatures shows that in the vicinity of the inner and outer strike points in hydrogen plasma the EEDF can be approximated by a bi-Maxwellian distribution, with a dominating low-energy electron population (4 - 7 eV) and a minority of electrons with higher energies (12 - 18 eV). In the private flux region between the two strike points, the EEDF is found to be Maxwellian with temperatures in the range of 7 - 9 eV. In the case of deuterium plasma under similar discharge conditions, the EEDFs in the vicinity of the inner and outer strike points, as well as in the private flux region between the two strike points, are found to be bi-Maxwellian.

1. Introduction

Langmuir probes (LPs) are relatively simple, inexpensive and robust diagnostic tools that are compatible with the tokamak scrape-off layer (SOL) conditions. They are readily deployable in large arrays built into the edge structures, such as limiters and divertor tiles. Because of their simplicity and the ability to perform local measurements of the electron energy distribution function (EEDF) and the plasma potential, LPs are frequently used in magnetized plasmas to characterize the edge plasma parameters [1]. Nevertheless, the interpretation of the electron part of the current-voltage (*IV*) characteristics above the floating potential remains a challenging subject in tokamaks because of the strong magnetic field effects.

This paper reports experimental data obtained on the COMPASS tokamak by Langmuir probes embedded in the divertor tiles [2]. The measured current-voltage probe characteristics were processed by means of the recently published [3,4] first-derivative probe technique (FDPT) for the precise determination of the plasma potential and the real EEDF, which can deviate from Maxwellian. The measurements were performed during L-mode hydrogen and deuterium plasma and the results are analyzed and discussed in the following sections.

2. The first-derivative Langmuir probe technique for evaluating the EEDF in tokamak edge plasma

The FDPT for evaluating the plasma parameters in tokamak edge plasma was published and discussed in detail in [3,4]. It was shown there that the electron current flowing to a cylindrical probe negatively biased by a potential U is given by:

$$I_e(U) = -\frac{8\pi e S}{3\mathrm{m}^2} \int_{eU}^{\infty} \frac{(W - eU) f(W) dW}{\gamma(W)\left[1 + \frac{(W - eU)}{W}\psi(W)\right]} \tag{1}$$

Here $f(\varepsilon)$ is the isotropic EEDF normalized by $\int_0^\infty f(\varepsilon)\sqrt{\varepsilon}d\varepsilon = \frac{4\pi\sqrt{2}}{m^{3/2}}\int_0^\infty f(W)\sqrt{W}dW = n_e$, where ε is the electron energy; e, m and n_e are the electron charge, mass and density; S is the probe area; U is the probe potential U_p with respect to the plasma potential U_{pl} $(U = U_p - U_{pl})$. The geometric factor γ assumes values in the range $0.71 \le \gamma \le 4/3$.

As was shown in [3] for probes oriented parallel to the magnetic field, the diffusion parameter is:

$$\psi^{\parallel}(W) = \frac{\pi L'}{64\gamma R_L(W,B)} \tag{2}$$

Here L' and $R_L(W,B)$ are the characteristic length of the turbulences and the electron Larmor radius. Taking into account (2) for the EEDF measured by a cylindrical probe parallel to the magnetic field, we obtain:

$$f(\varepsilon) = -\frac{3\pi\sqrt{2m}L'}{128e^3 S R_L(\varepsilon,B)U}\frac{\mathrm{d}I}{\mathrm{d}U}. \tag{3}$$

3. Langmuir probe measurements in the COMPASS divertor area

The evaluation of the plasma potential and EEDF was performed in a series of reproducible ohmic discharges in hydrogen and deuterium plasma. The results presented in this section are from typical COMPASS discharges labeled shot # 3908 (hydrogen plasma) and # 6036 (deuterium plasma) where the toroidal magnetic field was $B_T = 1.15$ T, the plasma current $I_p = 180$ kA and the average electron density $n_e = 8\times10^{19}$ m^{-3}. The divertor probe system in COMPASS consists of 39 single graphite Langmuir probes poloidally embedded in the divertor tiles; they are oriented parallel to the magnetic field and provide profiles with a spatial resolution in the poloidal direction down to 5 mm [5]. Each probe has an area of $S = 5{,}6\times10^{-7}$ m^2. These probes were electrically biased with respect to the tokamak chamber walls by means of triangular signal with a frequency of 1 kHz [6].

Results of the evaluation of the poloidal distribution of the plasma potential during hydrogen (# 3908) and deuterium (# 6036) discharges are presented in the figures 1a and 1b. The positions of the strike points are indicated by the vertical dashed lines: inner (Langmuir probe (LP) #7 at radial position $R = 0.419$ m) and outer (LP #24, $R = 0.501$ m) strike points for shot # 3908, and (LP #10, $R = 0.431$ m) and (LP#19, $R = 0.477$ m), respectively, for shot # 6036. During discharge # 6036 in deuterium, the X-point was closer to the divertor and, thus, strike points were closer to each other. It is seen that the plasma potential values in both shots decrease monotonically from the high-field side to the outer strike point and then increase (more pronounced for discharge # 6036).

a) b)

Figure 1. Poloidal profiles of the plasma potential obtained by the FDPT:
a) shot # 3908 in hydrogen and **b)** shot #6036 in deuterium.

The poloidal profile of electron temperatures in figure 2a shows that in hydrogen plasma in the vicinity of the inner and outer strike points the EEDF can be approximated by a bi-Maxwellian energy distribution. The low temperature (4 - 7 eV) population is dominant with presence of a small high energy (12 - 18 eV) electrons group. In the private flux region between the two strike points, the electron energy distribution function is found to be Maxwellian with temperatures in the range of 7 - 9 eV [3]. The corresponding densities are presented in figure 2b with the same symbols. We should point out that further theoretical and experimental investigations are needed to clarify the origin of the low-energy fraction in the bi-Maxwellian EEDF [7,8]. It should be noted that a bi-Maxwellian EEDF was also recently observed in the liquid lithium divertor area of National Spherical Torus Experiment (NSTX) at Princeton Plasma Physics Laboratory [9]. In these articles [7,8,9] a hypothesis accounting for the inelastic collision effects (i.e. excitation and ionization of neutral hydrogen or deuterium) is proposed to explain this EEDF feature.

The poloidal profile of the electron temperatures and densities corresponding to shot # 6036 in deuterium plasma are shown in figures 2c and 2d. The EEDF is bi-Maxwellian in the entire divertor region of this deuterium plasma.

Shot #3908, hydrogen **Shot #6036, deuterium**

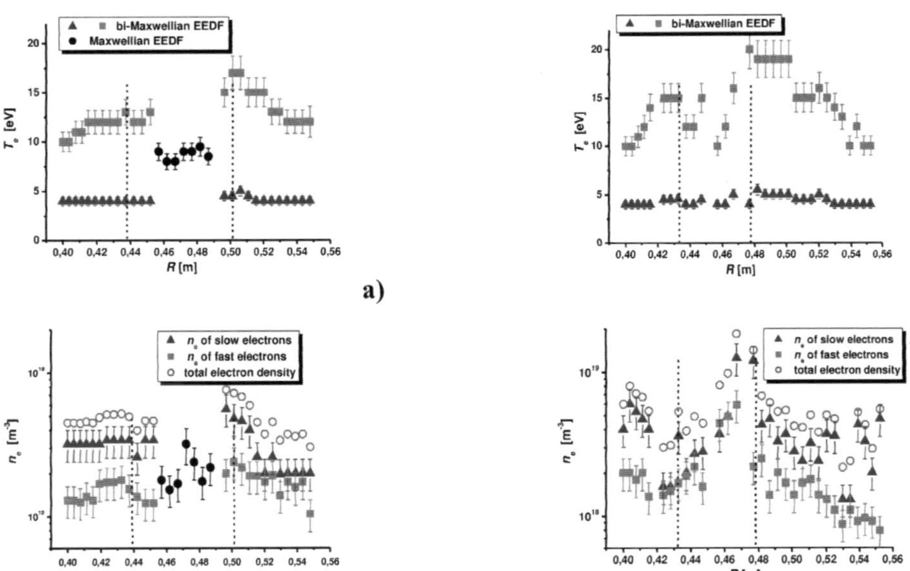

b) d)

Figure 2. Poloidal profiles of the electron temperatures and densities obtained by the FDPT for shot # 3908 (a and b) and # 6036 (c and d). The solid squares and triangles indicate the high and low temperatures and densities; the open circles are the temperatures and densities of the Maxwellian EEDF.

In the hydrogen discharge, the electron densities of the low and high energy fractions differ in the high field side (HFS) by a factor of ~3, and on the low field side (LFS), by a factor of ~1.5. In deuterium, the difference on HFS is ~4, around the outer strike point, ~2, and after radial position 0.52 m ~5.

4. Conclusions

This paper reports experimental data obtained on the COMPASS tokamak by Langmuir probes embedded in the divertor tiles. The first-derivative probe method was used for precise evaluation of the poloidal distribution of the plasma potential and the EEDF (respectively, the electron temperatures and the electron densities) during the current shots in hydrogen and deuterium ohmic L-mode plasma.

The spatial profile of the electron temperatures shows that in hydrogen plasma in the vicinity of the inner and outer strike points the EEDF can be approximated by a bi-Maxwellian distribution, with a dominating low-energy electron population and a minority of electrons with higher energies. In the private flux region between the two strike points, the electron energy distribution function is found to be Maxwellian.

In the case of deuterium plasma at similar conditions of the discharge, the EEDFs in the vicinity of the inner and outer strike points, as well as in the private flux region between the two strike points, are found to be bi-Maxwellian.

Acknowledgements

This research has been supported by the European Community under the contract of Association between EURATOM/IPP.CZ and EURATOM/INRNE.BG; by Bulgarian National Science Fund through the Association EURATOM-INRNE and by DNTS 01-10/28.11.2011, International Atomic Energy Agency (IAEA) Research Contract No 17125/R0, R1 as a part of the IAEA CRP F13014 on "Utilisation of a Network of Small Magnetic Confinement Fusion Devices for Mainstream Fusion Research" and 5th and 6th IAEA Joint Experiment, by the JOINT RESEARCH PROJECT between Institute of Plasma Physics v.v.i., AS CR and Institute of Electronics BAS BG, the grant project GA CR P205/11/2341, by MSMT Project # LM2011021 and project BG051PO001-3.3.05-0001 "Science and business".

References

[1] Popov Tsv K, Mitov M, Bankova A, Ivanova P, Dimitrova M, Rupnik S, Kovacic J, Gyergyek T, Cercek M, and Dias F M 2012 *Plasma Sources Sci. Technol.* **21**, 025004

[2] Pánek R, Bilyková P, Fuchs V, Hron M, Chráska P, Pavlo P, Stockel J, Urban J, Weinzettl V, Zajac J and Zacek F 2006 *Czechoslovak J. Phys.* **56**/2 B125–137

[3] Popov Tsv K, Ivanova P I, Stockel J and Dejarnac R 2009 *Plasma Phys. Control. Fusion* **51** 065014

[4] Dimitrova M, Dejarnac R, Popov Tsv K, Ivanova P, Vasileva E, Kovacic J, Stockel J, Havlicek J, Janky F and Panek R 2014 *Contributions of Plasma Physics* **54** No. 3 255 – 260

[5] Ivanova P, Popov Tsv K, Dimitrova M, Benova E, Bogdanov T, Stöckel J and Dejarnac R 2011 *Acta Technica* **56** T71-T83

[6] Mitov M, Bankova A, Dimitrova M, Ivanova P, Tutulkov K, Djermanova N, Dejarnac R, Stockel J and Popov Tsv K 2012 *Journal of Physics: Conference Series* **356**, 012008

[7] Tskhakaya D, Jachmich S, Eich T, Fundamenski W and JET EFDA Contributors 2011 *J. Nuclear Mater.* **415** S860-4

[8] Chodura R 1992 *Contr. Plasma Phys.* **32** 219-30

[9] Jaworski M A., Bell M G,. Gray T K, Kaita R, Kallman J, Kugel H W, LeBlanc B, McLean A G, Sabbagh S A, Soukhanovskii V A, Stotler D P, Surla V 2012 *Fusion Engineering and Design* **87** 1711– 1718

Journal of Physics: Conference Series **768** (2016) 012004 doi:10.1088/1742-6596/768/1/012004

Integral propagator method as a kinetic operator to describe discontinuous plasmas

J. M. Donoso, A. Jimenez , J. Gonzalez and L. Conde

Department of Applied Physics, Universidad Politecnica de Madrid, Madrid, Spain

E-mail: josemanuel.donoso@upm.es

Abstract.

We present a path-integral-like method to numerically solve drift-diffusion equations for plasma physics. The algorithm uses short-time propagators as approximate Green's functions that tend to smooth typical discontinuities arising in plasma dynamics as, for instance, the effects of plasma-wall interaction or localized particle flows. The usual numerical schemes based on differences may fail to represent these abrupt conditions by inducing numerical viscosities and instabilities. However, the robust mesh-free computational integral method has been proved to be unconditionally stable in the simple case of no imposed restrictions by boundary conditions. The extension of the method to deal with boundary value problems is analysed. The advancing scheme is also useful to deal with the merging of natural discontinuities in the system as those induced by the effects of electromagnetic fields generated by charge separation as well as for the existence of two differentiated plasma regimes. In any case, the kinetic equation may have drift and/or diffusion coefficients that are likewise discontinuous. The method works almost being numerically insensitive to these discontinuities, leading to feasible physically meaningful solutions. The scheme works as an effective integral kinetic operator even for Fokker-Planck equations connecting two different dynamical statistics, like Maxwell-Boltzmann and Fermi-Dirac ones. It is robust to describe spatially heterogeneous plasmas.

1. Introduction

As fluids, plasmas exhibit inherent discontinuous properties such as the species densities and localized currents or sheet structures. These configurations modify the global system behaviour entering into plasma description even in short time and spatial scales. The discontinuous quality of many plasma systems obeys to the their natural tendency in self-organizing into filamentary or layered charged structures. This property poses serious problems to numerically solve the associated model equations under given spatial and time scales. To analyse these natural discontinuities it is desirable to have robust stable algorithms able to localize and describe charge structures responsible of many phenomena. Aside this mentioned filamentary or layered structures, the interaction of a plasma with the containing walls is also an important topic that deserves much attention because the wall effects strictly drive time and steady states in laboratory plasmas. This plasma-wall yields to charge structures modifying the electrostatic potential profiles with somewhat non-continuous electric fields, for instance [1, 2]. In this sense, a heated emitting conductor surface constitutes in practice a diffuse boundary condition up to certain spatial scale that can be understood as a phase transition between mutually flowing electrons and ions. Thus, the surfaces of collecting and emissive walls constitute themselves a kind of discontinuity responsible of electric currents, charge separation and energy transport.

Within this natural scenario, integral propagator method for drift-diffusion differential equations is presented to analyze discontinuous plasma formations. The method is similar to that used by Chandrasekhar in his pioneer paper [3] and it has a solid modern mathematical background based upon Green's function theory and Markovian processes [4]. This path-integral approach has been applied to numerous problems in many scientific areas and has attracted much attention as a mesh-free method [5, 6, 7, 8]. To deal with these mentioned abrupt conditions, numerical schemes based upon differences may be effectiveness to properly model such a kind of behaviours without resorting to artificial smoothing functions that may disguise some physics. In some previous works we have applied the path-sum or integral method to non-linear plasma kinetic equations to deal with abrupt initial conditions [9, 10]. The smoothing effects and physically meaningful relaxation of any abrupt condition, as localized and collimated particle beams, has been effectively proved with these integral schemes. The integral method, also called Green's function method in other fields, makes use of an approximate short-time propagator suitable to establish a recurrence time advancing scheme to solve the problem.

The aim of this work is twofold, briefly summarizing some recent advances that improves and extends the use of the propagator integral method and introducing the method to address with discontinuous systems, or to the merging of natural discontinuities in plasmas. To achieve our first task we introduce the mathematical foundation of the propagator method accounting for the effects of source-sink terms in drift-diffusion equations and addressing a way to treat boundary value problems. On the other hand, the analysis of some introductory cases of Brownian systems with discontinuous deterministic and fluctuating forces leads us to properly substantiate the surprisingly quality of the propagator method to solve convection-diffusion problems with non-continuous coefficients. This property results particularly useful with regard to systems consisting in several quite different plasma scenarios separated by a physical interface or,even by two distinguishable physical behaviours described by different statistics, such as Fermi-Dirac and Boltzmann-Gibbs ones. In all cases, the path-integral-sum algorithm produces solutions that work as they were analytical

2. Mathematical Grounds

As referred above, many plasma processes obey integro-differential equations that can be cast into a drift-diffusion form as a generalized Fokker-Planck equation (FPE) , very common in multidisciplinary fields of mathematical physics [4, 11]. In plasma theory this form appears in fluid and kinetic descriptions to describe the time evolution of several plasma species fluid properties or the distribution function. If $\mathbf{q}$ represents a point of a $2N$ dimensional space, particularity a general six-dimensional point $(\mathbf{x}, \mathbf{v})$ of phase space, a general non-homogeneous (maybe non-linear) FPE for a distribution function $f = f(\mathbf{q}, t)$ shows the form

$$\frac{\partial f}{\partial t} = -\frac{\partial}{\partial \mathbf{q}} \cdot \left(\mathcal{A}\left(\mathbf{q}, t\,; f\right) - \frac{\partial}{\partial \mathbf{q}} \cdot \mathbb{D}(\mathbf{q}, t\,; f) \right) f(\mathbf{q}, t) + \rho(\mathbf{q}, t\,; f) = \mathbf{L}\, f + \rho \qquad (1)$$

where the operator $\mathbf{L}$ includes the symmetric diffusion tensor $\mathbb{D}$ and the drift vector $\mathcal{A}$, both may be non-linearly dependent on f. In conservative form $\mathbf{L}f$ can be rewritten as the divergence of a current $\mathbf{L}f = (\partial/\partial\mathbf{q}) \cdot \mathbf{J}$, where the vector $\mathbf{J}(\mathbf{q}, t) = \mathcal{A}f + \frac{\partial}{\partial \mathbf{q}} \cdot \mathbb{D}f$ is a flux called "density current" in statistical theory. The inhomogeneous contribution ρ stands for possible source-sinks terms. In particular, for the Vlasov-Fokker-Planck equation (VFPE) in plasma physics $f = f_a$ is the distribution function for species a. In this work we deal with a resolution method based on a semi-analytical approach combined with numerical computation, using propagators that are valid in the short-time regime of evolution. This method, also referred as path-sum, path-integral or integral numerical method in theoretically anchored in the well-known theory of Green's functions for linear parabolic differential equations. This solid mathematical theory is extrapolated to non-linear cases taken, among others, as a basis the theory of stochastic processes. In general, the f time evolution is rendered by means of a propagator $\Pi = \Pi(\mathbf{q}, t \mid \mathbf{q}', t')$ as

$$f(\mathbf{q}, t) = \int f(\mathbf{q}', t')\Pi(\mathbf{q}, t \mid \mathbf{q}', t')d\mathbf{q}' + \int_{t'}^{t} ds' \int d\mathbf{q}\ \rho(\mathbf{q}', s)\Pi(\mathbf{q}, t \mid \mathbf{q}', s) + f_B \qquad (2)$$

For a known Π that also satisfies (1) for an impulsive inhomogeneous term $\rho = \delta(\mathbf{q} - \mathbf{q}')\ \delta(t - t')$, or the homogeneous problem $\rho = 0$ with the initial Dirac-delta condition $\Pi(\mathbf{q}, t \mid \mathbf{q}', t) = \delta(\mathbf{q} - \mathbf{q}')$, the solution would be given for any initial condition $f(\mathbf{q}', t')$. The term f_B accounts for the effects of possible boundary conditions imposed to f in the advancing scheme that, for numerical implementation computing the f transition from time t to time $t+\tau$ can be rewritten in the more comprehensible alternate form

$$f(\mathbf{q}, t+\tau) = \int f(\mathbf{q}', t)\Pi(\mathbf{q}, t+\tau|\mathbf{q}', t)d\mathbf{q}' + \int_0^{\tau} ds \int d\mathbf{q}'\rho(\mathbf{q}', t+s)\Pi(\mathbf{q}, t+\tau|\mathbf{q}', t+s) + f_B. \quad (3)$$

Unfortunately, a generalized Green's function Π cannot be obtained if the original problem is analytically unsolvable, what enforces one to find an approximate propagator $P_\tau \approx \Pi$ valid for a short-time evolution $\tau > 0$, in order to apply (2) as an advancing scheme. This task can be made by seeking numerical Green's functions, as in [5], or using an analytical expression to build up a discrete propagator for numerical advancing the recurrence relation (3). Although this approximate short-time propagator P_τ is not unique [4, 11], in practice, its most usual form as a multi-variate Gaussian distribution

$$\Pi(\mathbf{q}, t+\tau; \mathbf{q}', t) \simeq P_\tau(\mathbf{q}, \mathbf{q}', t) = \exp\left(-\mathbf{Q}^T \cdot \mathbb{D}'^{-1} \cdot \mathbf{Q}\ /4\tau\right)\ /\ \parallel 4\pi\tau\mathbb{D}' \parallel^{1/2}. \qquad (4)$$

is efficient enough in open space problems with natural boundary conditions for numerical purposes. Here $\mathbf{Q} = \mathbf{q} - \mathbf{q}' - \tau\mathcal{A}'$ (as a column vector with transpose $\mathbf{Q}^T$). The diffusion tensor is assumed regular, namely $\parallel \mathbb{D} \parallel \neq 0$. This propagator, however, can serve as a simple basis to construct many others by adding second order corrections in τ to its mean vector and the covariance tensor if it is imperative to attain some usual conservative properties in many physical problems. Alternatively, noting that (4) is the fundamental solution of the homogeneous (1) for constant coefficients, following [12], it is possible to construct a simple auxiliary FPE , with τ as time variable, to find other useful expressions for some problems. In the advancing scheme, primes indicate functions computed at time $t' = t$ in the source or pre-point $\mathbf{q}'$ coordinates, instead of being evaluated in post-points or field points $\mathbf{q}$ at time $t + \tau$. Observe that for an initial Dirac-delta condition, $f(\mathbf{q}, t') = \delta(\mathbf{q} - \mathbf{q}_0)$, the scheme (2) smooths the solution from the very first time step. This feature holds for any abrupt discontinuous condition, a feature we claim to support the efficiency of the propagator method to solve drift-diffusion problems in physics.

3. Path-sum-boundary Method

One of the main difficulties usually addressed to apply the propagator method is related to the problem of dealing with boundary conditions in bounded spaces. This is still an open problem under research, even for the path-integrals in Quantum Mechanics [13]. A first theoretical approach to this respect was given in the pioneer work [14] with a rigorous discussion on how to solve the boundary value problem with (3). The additional boundary effects contribution f_B can be incorporated to the time advancing model by applying the general formula given in this work, which for slowly varying coefficients reads

$$f_B(\mathbf{q}, t) = -\int_{t'}^{t} ds \oint d\mathbf{S}' \cdot \left(\mathbf{J}(\mathbf{q}', s) + f(\mathbf{q}', s)\,\mathbb{D}(\mathbf{q}', s) \cdot \frac{\partial}{\partial \mathbf{q}'}\right)\Pi(\mathbf{q}, t \mid \mathbf{q}', s) \qquad (5)$$

This contribution is quite similar to the term associated to ρ in the generic integral scheme, as in electromagnetism, where boundary effects may be considered as they were source-sink terms over the surface $\mathbf{S}$. In a previous work we have applied a slightly different version of (5) that only takes into account the flux $\mathbf{J}$ at the surface $\mathbf{S}$ as

$$f_B = -\int_0^\tau ds \oint \Pi(\mathbf{q}, t+\tau|\mathbf{q}', t+s) d\mathbf{S}' \cdot \mathbf{J}(\mathbf{q}', t+s) \simeq -\int_0^\tau ds \oint P_{\tau-s}(\mathbf{q}, \mathbf{q}', t) d\mathbf{S}' \cdot \mathbf{J}(\mathbf{q}', t). \quad (6)$$

where the short-time propagator satisfies the homogeneous boundary condition of zero associated current at $\mathbf{S}$. The vector $\mathbf{J}$ can be prescribed or constructed for a particular problem with specified Dirichlet, Neumann or mixed boundary conditions as we explain in [10]. For the numerical implementation, the non-specified surface values of f and/or its derivatives can be replaced, for instance, by their values at previous time t. The term f_B can be simplified with a suitable choice of P_τ. To this task, one can construct a homogeneous auxiliary Fokker-Planck equation for $\partial P_\tau/\partial\tau$ dealing A and D as constant coefficients and using τ as the time variable. This problem can be solved to directly satisfy the boundary conditions for f of the real problem. The solution of this auxiliary problem can be obtained by standard techniques available in the literature or taken from the literature. As an example we refer here to the reflecting boundary condition problem (zero flux at the origin) in the half-line $(0, \infty)$. This imposes $J(P) = AP_\tau - \frac{\partial}{\partial q} DP_\tau = 0$, giving the alternate reflecting short-time propagator

$$P_\tau = \frac{e^{-Q_-^2}}{\sqrt{4\pi D'\tau}} + e^{\frac{A'}{D'}q} \left(\frac{e^{-Q_+^2}}{\sqrt{4\pi D'\tau}} - \frac{A'}{2D'} (1 - \mathrm{erf}\, Q_+) \right) \quad (7)$$

where $Q_\pm = (q \pm (q' + A'\tau))/\sqrt{4D'\tau}$ and erf is the error function. The effects of reflecting boundary conditions are provided by the last term, dominant for q close to the origin and responsible of accounting for reflecting conditions. With this P_τ the non-homogeneous boundary term, a given $J(0, t)$ for f, can be included in f_B by means of Eq. (6).

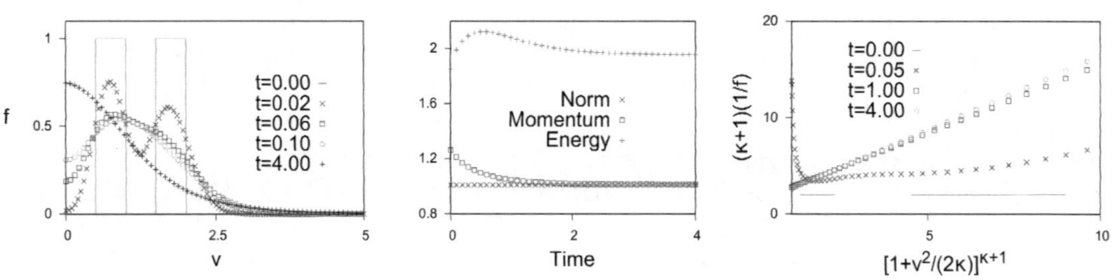

Figure 1. Evolution of f and its first three moments from an initial histogram-like with non-linear drift and power-law diffusion for (8) in dimensionless form, $\kappa = 4/3$.

Although P_τ is mainly controlled by a Gaussian distribution, the numerical adaptation of Eq. (2) with Eq. (4) does not imply that f behaves as a Gaussian. As an example, we set here a non-linear FPE with variable diffusion that leads to a power-law behaviour of the distribution tails. In particular for $q = v \geq 0$, we consider

$$A = -\omega\, v\, (1 - \Theta f) \quad , D = \omega \frac{kT}{m} \left(1 + \frac{m\, v^2}{2\, kT} \frac{1}{\kappa}\right) \quad (8)$$

where all the parameterr (written in a standard notation) can be set to unity. The parameter Θ can be related to quantum corrections due to possible particle degeneracy [15, 16, 17] used to model collision operators, it is related to the Planck's constant and particle mass ratio h/m. The dimensionless steady state solution is

$$f_s = \frac{\kappa+1}{\kappa}\left[1 + C\left(1+\frac{v^2}{2\kappa}\right)^{\kappa+1}\right]^{-1} \tag{9}$$

where C is an integration constant. This f_s which we have named Fermi-kappa distribution, tends to the usual Fermi-Dirac distribution for $\kappa \to \infty$, moreover, it turns to a Maxwellian for $C \gg 1$. Using the reflecting boundary conditions at the origin $v = 0$ with (7) as propagator, the time evolution is shown in figure 1. The numerical steady state solution results undistinguishable from the analytical one. Moments evolution are also shown, as well as a representation from which it can be inferred C, from the stationary straight line slope of the third frame.

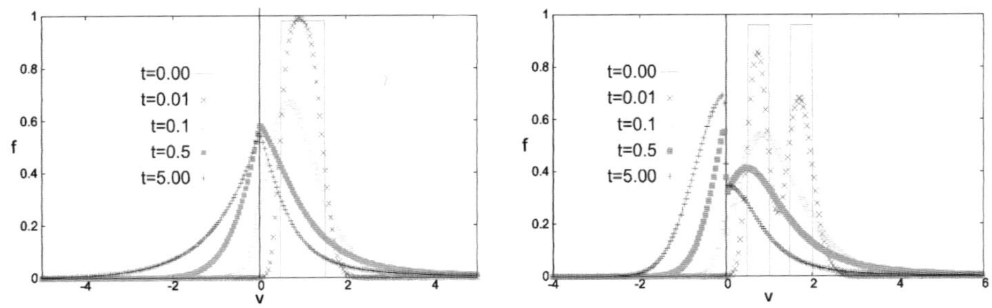

Figure 2. Evolution of f for the dry-friction Brownian motion and for discontinuous drift-diffusion functions (16).

4. Plasmas as Brownian Systems

By virtue of our experience in applying the propagator method, we can assert that considering the drift and diffusion coefficients computed in source (primed) variables reasonably works. We mean that the solution evolves as if $P_\tau(\mathbf{q}, \mathbf{q}'|t)$ were the unknown analytical propagator, that would include the correct dependence at any instant t without explicitly showing these coefficients. At this stage, we wonder if, as formal practical sources for the time evolution scheme, the coefficients A and D could be also non-continuous functions. Therefore, if these functions are understood as source terms, we have to check if P_τ could be valid for numerical implementation of such a kind of problems. This discontinuities are associated to abrupt changes of both dynamic (drift terms) and fluctuating forces (diffusion tensor) in a statistical interpretation of the system. We show here several cases that support the feasibility of the method to represent possible discontinuous forces, usually appearing in spatially heterogeneous systems that in plasmas can be understood as a generalized Brownian system described by

$$\frac{\partial f}{\partial t} + \underbrace{\mathbf{v}\cdot\frac{\partial f}{\partial \mathbf{x}}}_{spatial\,inhomogen.} + \underbrace{\mathbf{a}\cdot\frac{\partial f}{\partial \mathbf{v}}}_{external\,forces} = \underbrace{\left(\frac{\partial f}{\partial t}\right)_{coll} = \frac{\partial}{\partial \mathbf{v}}\cdot\left[\mathbf{D}_v - \frac{\partial}{\partial \mathbf{v}}\cdot\mathbf{D}_{vv}\right] f}_{Collisions} \tag{10}$$

where the deterministic (Newtonian) acceleration $\mathbf{a}$ usually stands for the electromagnetic forces. The effects of spatially inhomogeneities and collisions (in velocity space) have been also included. This equation is usually solved coupled to Poisson or Maxwell equations to self-consistently describe the nonlinear dynamics of the distribution function f [11].

Although most plasmas can be kinetically described by the collision-less Vlasov equation in the usual time and spatial scales, such a description should be accomplished by the collision

Journal of Physics: Conference Series **768** (2016) 012004 doi:10.1088/1742-6596/768/1/012004

operators, because the system can evolve in time to approach a non-Maxwellian distribution. Collisions always exist and they are, in fact, responsible of handing an equilibrium steady state. As in neutral gases, the best approximation for a kinetic collision operator is provided by the Boltzmann equation [15] for the so-called the binary approximation. A series of simplified collision operators can be derived from the Boltzmann model as, for instance, the Fokker-Planck-Landau one, which gives rise to a non-linear drift-diffusion equation [18, 19], whose numerical resolution is the aim of this work.

This Fokker-Plank-Landau collision operator in a multi species plasma accounts for the rate of change of the distribution function for species a of mass m_a, due to long-range Coulombian forces interactions, with field particles of species b (may be $a = b$) inside a Debye sphere, they both having a relative velocity $\mathbf{u} = \mathbf{v} - \mathbf{v}'$. The time evolution of f_a presents a contribution that can be expressed as the divergence in velocity space of a collision flux as

$$\left(\frac{\partial f_a}{\partial t}\right)_{coll.} = -\frac{\partial}{\partial \mathbf{v}} \cdot \sum_b \Gamma^{a/b} \int d\mathbf{v}' \mathbb{A}(\mathbf{u}) \cdot \left[\frac{f_a}{m_b}\frac{\partial f_b'}{\partial \mathbf{v}'} - \frac{f_b'}{m_a}\frac{\partial f_a}{\partial \mathbf{v}}\right] \tag{11}$$

where primes indicate that the corresponding functions are evaluated at $\mathbf{v}'$ whereas the second rank tensor $\mathbb{A}$ depend on $\mathbf{u}$ and can be connected to the dyadic unit dispersion tensor $(u^2\mathbb{I} - \mathbf{uu})/u^3$ with $u = |\mathbf{v} - \mathbf{v}'|$. The magnitudes $\Gamma^{a/b}$ are related to the characteristic plasma collision frequencies. This Fokker-Planck-Landau operator can be rearranged to the drift-diffusion form of (10) in velocity space [8, 9, 19]. Several simplified models can be obtained from (11), attending to the range of velocities where collisions are considered relevant in each particular system. Within the scope of this work we pay attention to this collision operator for two reasons. Firstly, we notice that a simple collision model accounting for quantum corrections (for fermions or bosons degeneracy accomplishing with the Pauli principle) can be easily included with the square brackets of Eq. (11) counter-changed as

$$\left[\frac{f_a}{m_b}\frac{\partial f_b'}{\partial \mathbf{v}'} - \frac{f_b'}{m_a}\frac{\partial f_a}{\partial \mathbf{v}}\right] \rightarrow \left[\frac{F_a}{m_b}\frac{\partial f_b'}{\partial \mathbf{v}'} - \frac{F_b'}{m_a}\frac{\partial f_a}{\partial \mathbf{v}}\right]. \tag{12}$$

The function $F(\mathbf{v}, t) = f(1-\Theta f)$ accounts for the quantum corrections in $\Theta = (h/m)^N \epsilon/\beta$ where h is the Planck's constant and the dimensionless parameter ϵ is 1, 0 or -1 (for Fermi-Dirac, Boltzmann and Bose-Einstein statistics respectively) and $\beta = 2$ for electrons. For spatially homogeneous conditions, a single species plasma approaches the same stationary solution given by the generalized collision Boltzmann equation

$$f_s = \beta \frac{m^N}{h^N} \frac{1}{C_0 \exp(\frac{mW^2}{2kT}) + \epsilon}, \quad \mathbf{W} = \mathbf{v} - \mathbf{u} \tag{13}$$

where C_0 is a normalization constant and $\mathbf{u}$ the mean velocity. If $C_0 >> 1$ no substantial deviation from the Maxwellian distribution (non-degenerate gas with $\epsilon = 0$) occurs and C_0 can be solved in terms of density n and temperature T.

On the other hand, we note that in (10) no cross x-v and x-x diffusion terms D_{xv}, D_{xx} appear, so that, the diffusion tensor is singular. However, this poses no problem, as proved in [20], because it can be used as a short-time propagator the solution of (10) with constant coefficients, giving a simple Gaussian probability density with means

$$\overline{v} = v' + A(v', x')\tau \quad \text{y} \quad \overline{x} = x' + v'\tau + \frac{1}{2}A(v', x')\tau^2,. \tag{14}$$

in the (1+1)-dim case. Here, $A = a + D_v$ is the total Newtonian acceleration of a test particle. Particularly, we have

$$P_\tau = \frac{\sqrt{3}}{2\pi\tau^2 D'} \exp\left(-\frac{1}{D'\tau}\left(3\frac{X^2}{\tau^2} - 3\frac{XV}{\tau} + V^2\right)\right) \tag{15}$$

with $X = x - \bar{x}$ and $V = v - \bar{v}$. Observe that both means coincide with the equations of deterministic motions in the short time approximation, up to order τ^3. Consequently, the covariance (depicting fluctuating effects) proves that effective cross x-v and self-diffusion xx processes physically appear in a slower time scale if compared to the v-v spreading. A finite-difference method would not properly account for these dispersive behaviours, because no information about x-x spatial diffusion is given in the differential equation (10).

5. Test Cases with Discontinuous Forces

To test the effectiveness of the propagator method, we firstly apply it to the analysis of the Brownian motion for dry friction force $F = -\mu\ sign(v)$, as alternative to the usual viscosity force $F = -\nu v$. This problem can be solved to obtain both analytical transient solution and non-Maxwellian steady state [21]. In particular for a spatially uniform system, we have tested that the integral method describes the f evolution as no discontinuous force exists, in a very good agreement with analytical results (first frame of Fig. 16) . Therefore, to explore the capabilities of the method we present here a case of non-linear A and piecewise defined non continuous D coefficient as

$$A = \begin{cases} 1 \quad if\, \mathrm{v} < 0 \\ -\mathrm{v}\ (1-f), if\ \mathrm{v} > 0 \end{cases} \qquad D = \begin{cases} 0.1 \quad if\, \mathrm{v} < 0 \\ 1 + \frac{\mathrm{v}^2}{2\kappa}\ , if\ \mathrm{v} > 0 \end{cases} \tag{16}$$

The results (second frame of Fig. 2) evidence that discontinuities in f are intrinsically correlated to the discontinuities in D. The changes in D are in connexion to the variable intensity of fluctuating forces (stochastic noise or collisions in plasmas). The analytical steady state solution f_s can be easily obtained, the numerical solution is in excellent agreement with f_s, whose form for $v > 0$ coincides with (9). This latter results suggests that the boundary problem dealt above can be also stated by extending f to the region $v < 0$.

6. The Two Statistics Connexion

The problem of a plasma bulk interacting with a heated emissive wall [1, 22] is studied as a boundary value problem. However, since a physical interface appears as continued inside the plasma to a certain scale, we propose to model the wall effect as a transition where the ensemble of electrons behaves as degenerate fermions up to some extent. This transition from a degenerate

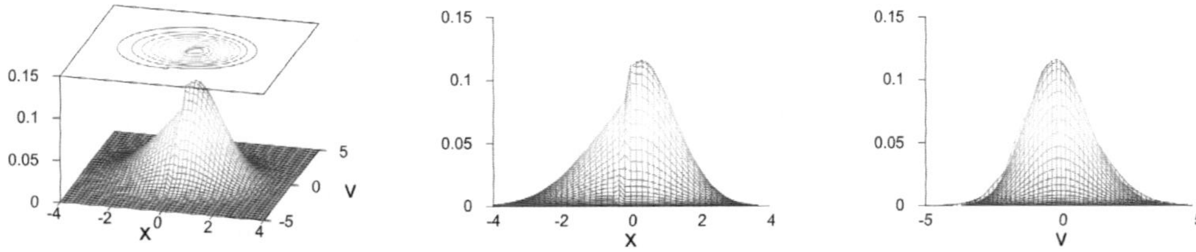

Figure 3. Distribution $f(x, v)$ with x and v projections for the two statistics plasma-wall connexion.

electron system in the material to a non-degenerate ensemble in the plasm bulk is, in fact, a discontinuity of the form we are studying. Hence, we consider the modified Fokker-Planck equation for Fermi-Dirac particles in order to apply the propagator integral method to such a transition with (15). The problem is then formulated as an heterogeneous plasma in whole phase space, in planar geometry, with piecewise defined kinetic coefficients. Moreover, the coupling with self-consistent Poisson equation establishes a treatment for which only wall electrostatic potential and bulk plasma neutrality governs the system dynamics. In resume, we consider

$$\frac{\partial f}{\partial t}+\mathrm{v}\frac{\partial f}{\partial x}+\frac{qE(x,t)}{m}\frac{\partial f}{\partial \mathrm{v}}=\begin{cases}-\frac{\partial}{\partial \mathrm{v}}\left[-\nu_w(\mathrm{v}-u)(1-\Theta f)-\nu_w\frac{kT_w}{m}\frac{\partial}{\partial \mathrm{v}}\right]f, x<0\\ -\frac{\partial}{\partial \mathrm{v}}\left[-\nu(\mathrm{v}-u)-\frac{\nu kT}{m}\frac{\partial}{\partial \mathrm{v}}\right]f,\ x>0\end{cases} \tag{17}$$

for different average collision frequencies ν and diffusion terms, as functions of wall and plasma temperatures T_w and T. The quantum correction is included in the material medium in order to simulate this contact as an interface or thin layer. Here $\Theta = h/2m$ and $u = u(x,t)$ is the fluid velocity. In dimensionless form, qualitative interesting results are obtained for the electron distribution function evolving in a neutralizing background of massive ions at rest. A first case, plotted in Fig. 3, where the initial condition is a Dirac-delta function (only one channel in the computational discrete domain). The resulting distribution f in xv space is not a Maxwellian nor a Fermi-Dirac one close the interface $x = 0$, it varies abruptly from $x < 0$ to $x > 0$ regions, whereas it is almost continuous in v. Any other initial condition would evolve as a superposition of solutions of this case. An interesting case takes as initial condition an uniform electron density $n = 1$ (dimensionless), and emission from the wall distorts the initial plasma quasi neutrality. The wall emits to a plasma in presence of an electric field that evolves, due to charge separation, to a steady state in which the electron flux is continuous and the plasma bulk remains neutral (Fig. 4).

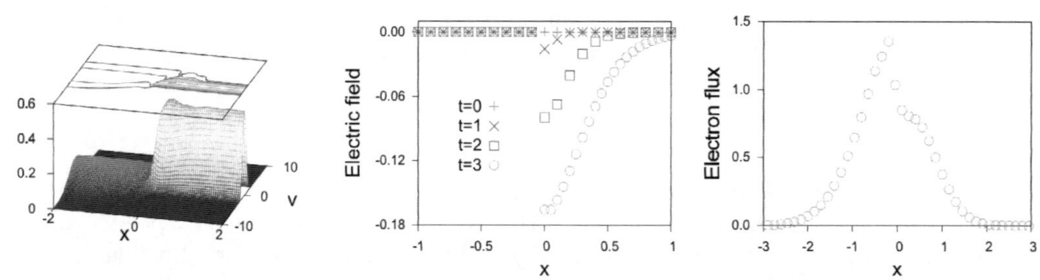

Figure 4. Distribution $f(x,v)$ for two-statistics electron plasma with $n(x) = 1$ at $t = 0$. The time evolving electric field and the steady electron flux are also presented.

7. Conclusions

In spite of the integral propagator method is supported by a rigorous mathematical basis, its capability as a computational method has not been emphasized enough till now. Furthermore, its certain resemblance to the path-integral formalism of Quantum Mechanics has not stimulated the method, mainly due to the handicap of how to deal with boundary value problems. However, some recent advances have been achieved to attract attention on the feasibility and advantages of the propagator integral method. Apart of the already well-established properties as, its capability of dealing with non-linear and non-homogeneous problems under disparate spatial and time scales of different plasma species, robustness, mesh-free scheme, etc. some other advantages have to be emphasized. The recursive time procedure makes the numerical solution to evolve as it would be with the exact unknown propagator. Smoothing effects of diffusion processes for drastic variations of the coefficients have been addressed for the first time. This property comes out of having considered the convection and spreading coefficients as sources that can be discontinuous and piecewise given, being this P_τ continuous with respect to field variables. The propagator method properly picks up the effects of natural heterogeneous variations of A_i and D_{ij}

that are very difficult to compute with finite difference schemes. In addition, some phenomena, as plasma-wall interaction, can be modelled considering plasma as heterogeneous media with drastic variations in the dynamic forces or by a transition between two subsystems described by different statistics. This interpretation can be also used to deal or establish physically meaningful boundary conditions. In kinetic theory, the time advancing scheme replaces the differential equation by an integral one that can be numerically implemented. This feature suggests that the propagator method acts, in fact, as a kinetic operator for any afforded drift-diffusion equation problem.

References

[1] Lieberman M A and Lichtenberg A J 1994 *Principles of plasma discharges and materials processing* (Cambridge Univ Press)
[2] Smirnov B M 2012 *Fundamentals of Ionized Gases: Basic Topics in Plasma Physics* (John Wiley & Sons)
[3] Chandrasekhar S 1943 *Rev. Mod. Phys.* **15**(1) 1–89
[4] Risken H 1989 *The Fokker-Planck Equation* (Berlin: Springer)
[5] Loureiro F, Mansur W, Wrobel L and Silva J 2014 *International Journal of Heat and Mass Transfer* **76** 393–404
[6] Kleinert H and Zatloukal V 2013 *Phys. Rev. E* **88**(5) 052106
[7] Xie W X, Xu W and Cai L 2008 *Applied Mathematics and Computation* **197** 426 – 434
[8] Hitchon W, Koch D and Adams J 1989 *Journal of Computational Physics* **83** 79 – 95
[9] Donoso J M, Salgado J J and Soler M 2005 *Journal of Physics A: Mathematical and General* **38** 9145
[10] Gonzalez J, Donoso J M and Conde L 2014 *Contributions to Plasma Physics* **54** 298–303
[11] Frank T D 2005 *Nonlinear Fokker-Planck equations: fundamentals and applications* (Springer)
[12] Donoso J M, Salgado J J and Soler M 1999 *Journal of Physics A: Mathematical and General* **32** 3681
[13] Asorey M, Clemente-Gallardo J and Munoz-Castaneda J 2007 *Journal of Physics: Conference Series* vol 87 (IOP Publishing) p 012004
[14] Wehner M F and Wolfer W G 1983 *Phys. Rev. A* **28**(5) 3003–3011
[15] Chapman S, Cowling T G and Burnett D 1970 *The mathematical theory of non-uniform gases* (Cambridge University Press)
[16] Hu J, Jin S and Yan B 2012 *Commun Comp Phys* **12** 1541–1561
[17] Liu S, Ma X and Yu H 2012 *Journal of Differential Equations* **252** 5414 – 5452
[18] Braginskii S 1965 *Reviews of plasma physics* **1** 205
[19] Trubnikov B 1965 *Reviews of plasma physics* **1** 105
[20] Donoso J M and del Rio E 2007 *Journal of Physics A: Mathematical and Theoretical* **40** F449
[21] Touchette H, Prellberg T and Just W 2012 *Journal of Physics A: Mathematical and Theoretical* **45** 395002
[22] Tierno S, Domenech-Garret J, Donoso J, Jennewein D, Herdrich G, Fasoulas S and Conde L 2013 *Plasma Science, IEEE Transactions on* **41** 695–700

Sheath formation in an oblique magnetic field – some comments on length scales and source terms

T Gyergyek [1,2], **J Kovačič**[2]
[1] University of Ljubljana, Faculty of electrical engineering, Tržaška 25, 1000 Ljubljana, Slovenia
[2] Jožef Stefan Institute, Jamova 39, 1000 Ljubljana, Slovenia

E-mail: tomaz.gyergyek@fe.uni-lj.si

Abstract. A one-dimensional fluid model is presented and used for the analysis of the potential formation in front of a negative, large, planar electrode immersed in magnetized plasma. The magnetic field lines can form an arbitrary angle with the electrode surface, with the exception that they must not be completely parallel to the surface. The model includes elastic ion collisions and 3 source terms that describe ionization mechanisms. If the exponential source term is used, the pre-sheath is longer than in the case of the constant source term. The zero source term results in qualitatively different solutions that may even exhibit nonphysical oscillations. The problem involves 4 characteristic length scales. Any of them can be used for the normalization of the space coordinate and this has no effect to the solutions.

1. Introduction

The boundary of magnetized plasma is of great interest to several fields ranging from material processing to magnetic confinement fusion. The scrape-off layer of a tokamak, for example has low density and strong magnetic fields. Further, the angle of incidence of the magnetic field lines onto the divertor spans nearly the whole range from perpendicular to parallel. In this paper we investigate the structure of the sheath and pre-sheath using a simple fluid description. Similar models [1]-[4] have been used to study the pre-sheath region only. In this work the sheath and the pre-sheath region are studied together. The presented model includes the effects of elastic collisions, finite ion temperature and three model of charged particle creation.

2. Model

Formation of a sheath in front of a negative planar electrode immersed in a magnetized plasma is studied by a one dimensional fluid [1,2] model. The ions are assumed to obey the continuity equation and equation of motion:

$$\frac{\partial n_i}{\partial t} + \nabla \cdot (n_i \mathbf{u}_i) = S_i, \tag{1}$$

$$m_i n_i \left(\frac{\partial \mathbf{u}_i}{\partial t} + (\mathbf{u}_i \cdot \nabla) \mathbf{u}_i \right) = n_i e_0 (\mathbf{E} + \mathbf{u}_i \times \mathbf{B}) - \nabla p_i + \mathbf{A}_i - m_i \mathbf{u}_i S_i. \tag{2}$$

Here m_i is the ion mass, t is time, $\mathbf{u}_i$ is the ion fluid velocity, n_i is the ion density, e_0 is the elementary charge, $\mathbf{E}$ is electric field, $\mathbf{B}$ is magnetic field, p_i is the ion pressure, S_i is the source term and $\mathbf{A}_i$ is the collision term. Let us first focus on the collision term $\mathbf{A}_i$. The ions can exchange momentum in elastic collisions with other particle species. In our model those species could be electrons and neutral atoms of the same kind as the ions. The rate of change of momentum of the ions because of such collisions should therefore be written in the following way:

$$\mathbf{A}_i = -m_i n_i \nu_{in}(\mathbf{u}_i - \mathbf{u}_n) - m_i n_i \nu_{ie}(\mathbf{u}_i - \mathbf{u}_e).$$

Here ν_{in} is the frequency of ion neutral collisions, ν_{ie} is the frequency of ion electron collisions , $\mathbf{u}_n$ is the neutral fluid velocity and $\mathbf{u}_e$ is the electron fluid velocity. Elastic collisions with neutrals could be charge exchange collisions or also of some other type. Collisions with the electrons on the other hand are coulomb collisions. It is beyond the scope of this work to analyse in detail the collisions of ions with other particle species. No attempt is made to estimate the relative ion velocity with respect to other particle species. Instead the collision term $\mathbf{A}_i$ is taken in the following simple form:

$$\mathbf{A}_i = -m_i n_0 \nu \mathbf{u}_i. \tag{3}$$

Here ν is the ion collision frequency and n_0 is the density of ions in the region where the plasma is not perturbed by the electrode, this means beyond the pre-sheath.

Three different source terms are considered. The first one is called *the zero source term*:

$$S_i = 0. \tag{4}$$

This means that no ions are produced or annihilated anywhere in the plasma. The second is the *constant source term:*

$$S_i = \frac{n_0}{\tau}. \tag{5}$$

Here τ is the so called ionization time. The source term S_i gives the difference between the number of created and the number of annihilated ions per unit volume and per unit time. The third source term that is considered in this work is the *exponential source term*:

$$S_i = \frac{n_0}{\tau}\exp\left(\frac{e_0\Phi}{kT_e}\right). \tag{6}$$

Here Φ is the potential, which is space dependent, k is the Boltzmann constant and T_e is the electron temperature.

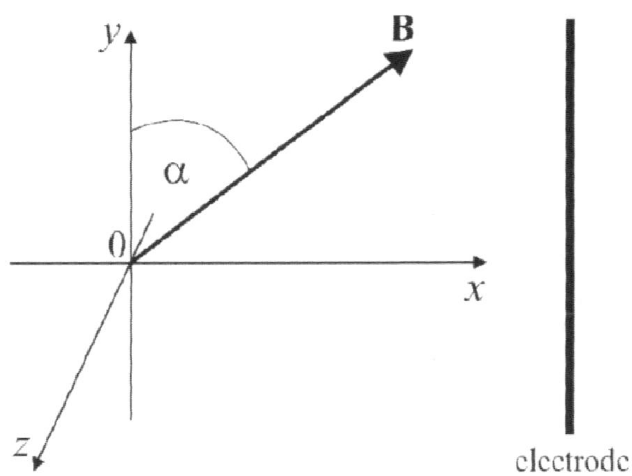

Figure 1. Schematic of the model. A large planar electrode is perpendicular to the x axis. The magnetic field lies in the xy plane. The magnetic field angle α is the angle between the y axis and the magnetic field vector **B**.

An infinitely large planar electrode is considered. The electrode is perpendicular to the x axis. The magnetic field lies in the xy plane, as shown in figure 1. The components of the magnetic field are therefore given by $\mathbf{B} = B(\sin\alpha, \cos\alpha, 0)$. Our model is one dimensional. This means that in our model the electric field only has one component: $\mathbf{E} = (E_x, 0, 0)$, which is given by:

$$E_x = -\frac{\mathrm{d}\Phi}{\mathrm{d}x}.$$

The gradient operator is replaced by the derivative over x:

$$\nabla \to \frac{\mathrm{d}}{\mathrm{d}x}.$$

The following equation of state is assumed for the ions:

$$p_i(x) = \kappa n_i(x) kT_i.$$

In a one dimensional model the pressure depends only on x coordinate. Also the ion temperature T_i could in principle be a function of x. But in our model we assume that the ion temperature is constant everywhere and consequently the polytropic coefficient κ is equal to 1, $\kappa = 1$. But in general κ can also be a function of x [5]. So the pressure gradient in equation (2) is replaced by

$$\nabla p_i \to kT_i \frac{\mathrm{d}n_i}{\mathrm{d}x}.$$

The electrons are assumed to be Boltzmann distributed with a given temperature T_e:

$$n_e(x) = n_0 \exp\left(\frac{e_0 \Phi(x)}{kT_e}\right).$$

Now also the idea behind the exponential source term becomes clear. The assumption behind the exponential source term is that the main ionization mechanism are ionizing collisions of electrons with neutral atoms.

In the steady state the time derivatives are zero, so the equations (1) and (2) become ordinary differential equations. In the coordinate system defined in figure 1 and using the one dimensional model, the equations (1) and (2) are written in the following way:

$$\frac{\mathrm{d}n}{\mathrm{d}X} V_x + n \frac{\mathrm{d}V_x}{\mathrm{d}X} = \frac{\mathrm{d}}{\mathrm{d}X}(nV_x) = A_1 \left\{ \begin{matrix} 0 \\ 1 \\ \exp(\Psi) \end{matrix} \right\}, \tag{7}$$

$$nV_x \frac{\mathrm{d}V_x}{\mathrm{d}X} + n \frac{\mathrm{d}\Psi}{\mathrm{d}X} + \Theta \frac{\mathrm{d}n}{\mathrm{d}X} + A_2 \cos\alpha\, nV_z + A_1 V_x \left\{ \begin{matrix} Z \\ 1+Z \\ \exp(\Psi)+Z \end{matrix} \right\} = 0, \tag{8}$$

$$nV_x \frac{\mathrm{d}V_y}{\mathrm{d}X} - A_2 \sin\alpha\, nV_z + A_1 V_y \left\{ \begin{matrix} Z \\ 1+Z \\ \exp(\Psi)+Z \end{matrix} \right\} = 0, \tag{9}$$

$$nV_x \frac{\mathrm{d}V_z}{\mathrm{d}X} - A_2 \cos\alpha\, nV_x + A_2 \sin\alpha\, nV_y + A_1 V_z \left\{ \begin{matrix} Z \\ 1+Z \\ \exp(\Psi)+Z \end{matrix} \right\} = 0. \tag{10}$$

The system in completed with the Poisson equation in one dimension, which determines the potential profile:

$$A_3 \frac{d^2\Psi}{dX^2} = \exp\Psi - n. \tag{11}$$

The following variables have been introduced:

$$\lambda_D = \sqrt{\frac{\varepsilon_0 kT_e}{n_0 e_0^2}},\ c_0 = \sqrt{\frac{kT_e}{m_i}},\ \omega_c = \frac{e_0 B}{m_i} = \frac{2\pi}{T_{cyc}},\ K = \omega_c \tau,\ L = c_0 \tau,\ \mu = \frac{m_e}{m_i},$$
$$L_c = \frac{c_0}{\nu},\ r_{Li} = \frac{c_0}{\omega_c},\ n = \frac{n_i}{n_0},\ \Psi = \frac{e_0\Phi}{kT_e},\ \Theta = \frac{T_i}{T_e},\ \varepsilon = \frac{\lambda_D}{L},\ Z = \nu\tau = \frac{\tau}{T_{coll}}, \tag{12}$$
$$V_x = \frac{u_x}{c_0},\ V_y = \frac{u_y}{c_0},\ V_z = \frac{u_z}{c_0},\ X = \frac{x}{\lambda_D} \text{ or } X = \frac{x}{L} \text{ or } X = \frac{x}{r_{Li}} \text{ or } X = \frac{x}{L_c}.$$

Here ε_0 is the permittivity of the free space and T_{cyc} is the cyclotron period. The potential Φ is normalized to the electron temperature divided by elementary charge kT_e/e_0. The components of the ion velocity are normalized to the so called normalizing velocity c_0, defined in (12). The problem is characterized by 4 characteristic length scales: the ionization length L, the Debye length λ_D, the mean free path or collision length L_c and the ion Larmor radius r_{Li}. The ionization length L is the distance that an ion moving with the normalizing velocity c_0 passes in one ionization time τ. The mean free path L_c is the distance that an ion moving with the normalizing velocity c_0 passes in the so called collision time $T_{coll} = 1/\nu$. An ion that enters the magnetic field B with normalizing velocity c_0 perpendicularly to the magnetic field line gyrates with the Larmor radius r_{Li}. Obviously the space coordinate x can be normalized to any of these lengths. From (12) it can be seen easily that:

$$K = \frac{L}{r_{Li}} = \frac{2\pi\tau}{T_{cyc}},\ Z = \frac{L}{L_c},\ \frac{K}{Z} = \frac{L_c}{r_{Li}} = \frac{2\pi T_{coll}}{T_{cyc}},\ K\varepsilon = \frac{\lambda_D}{r_{Li}},\ Z\varepsilon = \frac{\lambda_D}{L_c}. \tag{13}$$

A remark should be given about the ion sound velocity c_S. The ion sound velocity c_S is defined as:

$$c_S = \sqrt{\frac{kT_e^* + \kappa kT_i}{m_i}}. \tag{14}$$

Here T_e^* is the so called screening temperature [5]. In our case it is equal to T_e because the electrons are the only negatively biased particle species. Using (12) and taking into account that $\kappa = 1$ the ion sound velocity (14) is written as:

$$V_S = \frac{c_S}{c_0} = \sqrt{1+\Theta}.$$

The coefficients A_1, A_2 and A_3 depend on the normalization of the space coordinate x and they are given in table 1.

Table 1. The coefficients A_1, A_2 and A_3 for various normalizations of the space coordinate x.

Symbol	$X=x/L$	$X=x/\lambda_D$	$X=x/L_c$	$X=x/r_{Li}$
A_1	1	ε	$1/Z$	$1/K$
A_2	K	$K\varepsilon$	K/Z	1
A_3	ε^2	1	$Z^2\varepsilon^2$	$K^2\varepsilon^2$

The system (7) - (11) represents 12 (3 source terms × 4 normalizations of x) systems of 5 equations for 5 unknown functions of X, which are $n(X)$, $\Psi(X)$, $V_x(X)$, $V_y(X)$ and $V_z(X)$. Once these functions are found, one may find also the parallel velocity V_{par}, the perpendicular velocity V_{perp} and the angle of incidence β, which are given by:

$$V_{par} = \frac{\mathbf{V}\cdot\mathbf{B}}{B} = V_x \sin\alpha + V_y \cos\alpha,$$

$$V_{perp} = \frac{|\mathbf{V}\times\mathbf{B}|}{B} = \sqrt{V_z^2 + \left(V_x \cos\alpha - V_y \sin\alpha\right)^2},$$

$$\beta = \arctan\left(\frac{V_x}{V_y}\right).$$

Here parallel and perpendicular refers to the magnetic field.

3. Results

We now present some results of the model presented in the previous section. In order to find the unknown functions $n(X)$, $\Psi(X)$, $V_x(X)$, $V_y(X)$ and $V_z(X)$ from the system (7) - (11) first the normalization of x and the source term must be selected and then the parameters K, Z, ε, α and Θ must also be selected. In the next step boundary conditions must be selected. From figure 1 it can be seen that the numerical integration of the equations starts at $X = 0$ and proceeds in the positive X direction towards the electrode, so strictly mathematically speaking, one is dealing with initial value problem. The "initial" conditions (since X is the space coordinate these are in fact *boundary conditions*) that are usually selected are the following:

$$n(0)=1,\ \Psi(0)=0,\ \frac{\mathrm{d}\Psi}{\mathrm{d}X}(0)=0,\ V_x(0)=V_0,\ V_y(0)=0,\ V_z(0)=0. \tag{15}$$

It is assumed that at the starting point $X = 0$ the plasma is not perturbed, so the plasma density there is n_0, which gives the first boundary condition. In the unperturbed plasma also electric field must be zero and this gives the third boundary condition. The second boundary condition simply tells that the plasma potential at $X = 0$ is selected as the zero of the potential. In our model the ions are born at rest so all three velocity components should be zero. But in this case only the zero solution of the system can be found. So a small starting velocity [1,3,4] in the positive X direction must be selected. A typical value is $V_0 = 10^{-8}$. The system of equations is integrated using the fourth order Runge-Kutta method with variable step size. The size of the step is decreased only a limited number of times (called iterations), which is specified beforehand. If after the prescribed number of iterations the desired precision has not yet been reached, the system is declared to be stiff, since in the point, where this occurs there is very probably a singularity.

In figure 2 an example of the solution of the system (7) - (11) is shown. The following parameters are selected: $K = 50$, $Z = 2$, $\alpha = 20^\circ$, $\Theta = 0$ and 2 values of ε are selected, $\varepsilon = 10^{-3}$ and $\varepsilon = 10^{-5}$. The boundary conditions are given (15) with $V_0 = 10^{-8}$ and the exponential source term (6) is selected. The space coordinate is normalized to the Larmor radius r_{Li}. Solutions obtained with $\varepsilon = 10^{-3}$ are plotted with solid line while the solutions obtained with $\varepsilon = 10^{-5}$ are plotted with a dotted line. In the top left figure the velocity component $V_x(X)$ is shown. The ion sound velocity V_S is also labelled. The point where V_x reaches V_S is identified as the sheath edge. In the top middle graph the potential profile $\Psi(X)$ is presented. In the top right plot the electric field profile is displayed. In the bottom left figure the ion $n(X)$ and electron density $\exp(\Psi(X))$ profiles are shown. On the insert a part of the figure is shown on an expanded scale. In the bottom middle plot the space charge density profile is shown. This is simply the difference between the ion and electron density $n(X) - \exp(\Psi(X))$. Finally in the bottom right figure the angle of incidence $\beta(X)$ is plotted. Note that according to the boundary conditions, $\beta(0)$ should be 90 degrees. The starting point of the integration is eliminated from the plot. The ion flow follows the

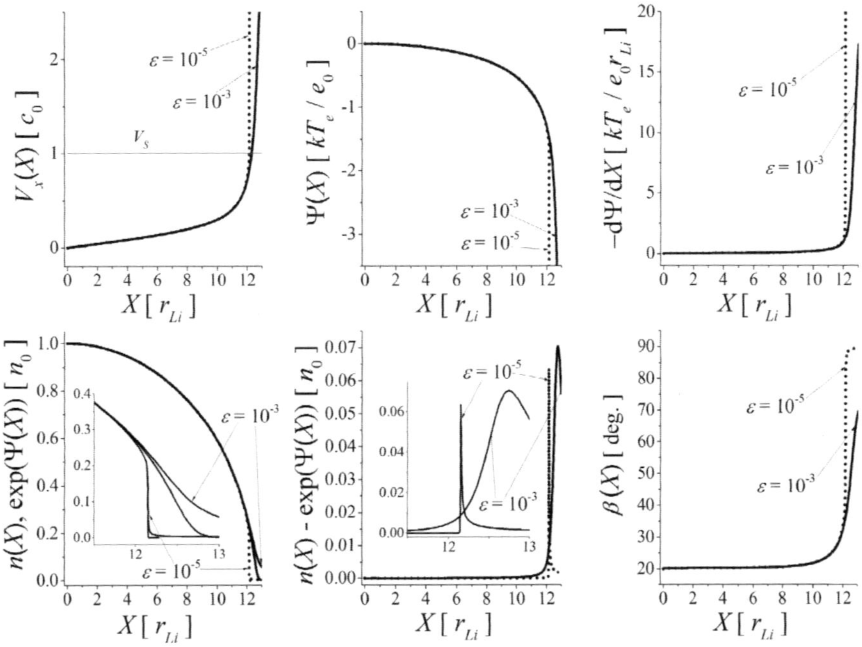

Figure 2. An example of the solution of the system (7) - (11). The parameters are: K = 50, Z = 2, α = 20°, Θ = 0 and 2 values of ε are selected, $\varepsilon = 10^{-3}$ (solid) and $\varepsilon = 10^{-5}$ (dotted). The boundary conditions are given by (15) with $V_0 = 10^{-8}$ and the exponential source term (6) is selected. The space coordinate is normalized to the Larmor radius r_{Li}.

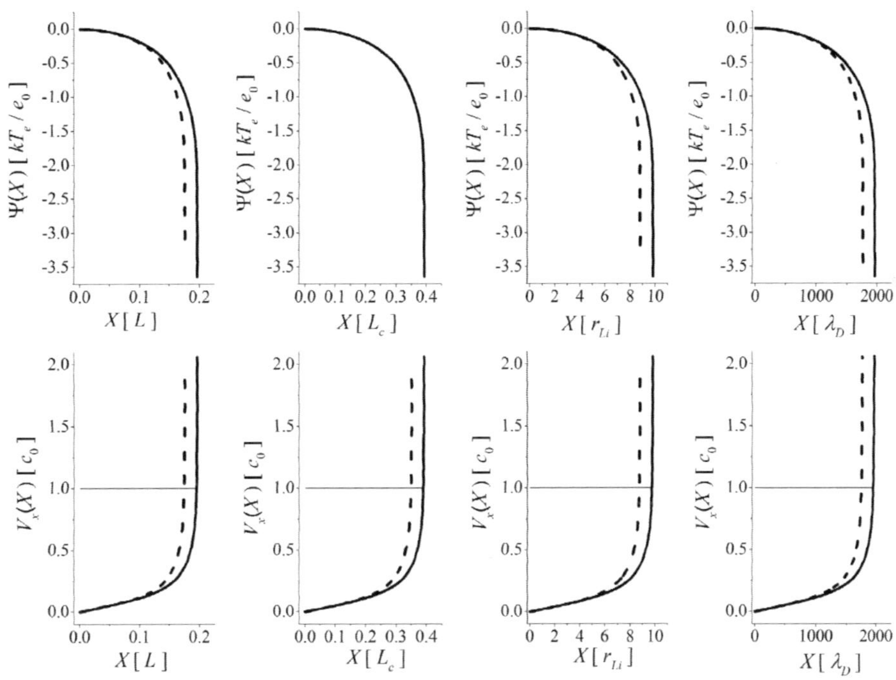

Figure 3. Potential $\Psi(X)$ (top) and velocity profiles $V_x(X)$ (bottom) found from the system (7) - (11) for the parameters K = 50, Z = 2, α = 20°, Θ = 0 and $\varepsilon = 10^{-4}$. The boundary conditions are given by (15) with $V_0 = 10^{-8}$. The exponential (solid) and the constant (dashed) source term are used. The space coordinate is normalized to all 4 characteristic lengths.

magnetic field lines almost to the sheath edge and then within approximately 1 Larmor radius it is diverted by the electric field into direction which is almost perpendicular to the electrode. The selection of ε has strong impact to the solutions especially close to the sheath edge.

In figure 3 properties of the solutions of the system (7) - (11) are illustrated further. The following parameters are selected: $K = 50$, $Z = 2$, $\alpha = 20^{\circ}$, $\Theta = 0$ and $\varepsilon = 10^{-4}$. The boundary conditions are given by (15) with $V_0 = 10^{-8}$. The constant and the exponential source term are selected. The solutions obtained with the exponential source term are shown with solid line, while the solutions that correspond to the constant source term are dashed. The space coordinate is normalized to all 4 characteristic lengths. The potential profiles $\Psi(X)$ are shown in the top graphs, while the velocity profiles $V_x(X)$ are displayed in the bottom plots for all space coordinate normalizations. From the left to the right the normalizing lengths go as follows: the ionization length L, the mean free path L_c, the Larmor radius r_{Li} and the Debye length λ_D. It can be seen clearly that the normalization of the space coordinate has no effect to the solutions, except that the horizontal scale is expanded or stretched, as prescribed by the values of K, Z and ε. Note that one ionization length correspond to 2 mean free paths and to 50 Larmor radii and to 10000 Debye lengths, exactly as determined by $K = 50$, $Z = 2$ and $\varepsilon = 10^{-4}$.

In figure 4 very similar solutions of the system (7) - (11) are shown as in figure 3. Even the same parameters are selected: $K = 50$, $Z = 2$, $\alpha = 20^{\circ}$, $\Theta = 0$ and $\varepsilon = 10^{-4}$. Only this time the zero source term is used. This time V_0 must be increased for 7 orders of magnitude in order to obtain the solutions with similar pre-sheath lengths as with the exponential and constant source terms.

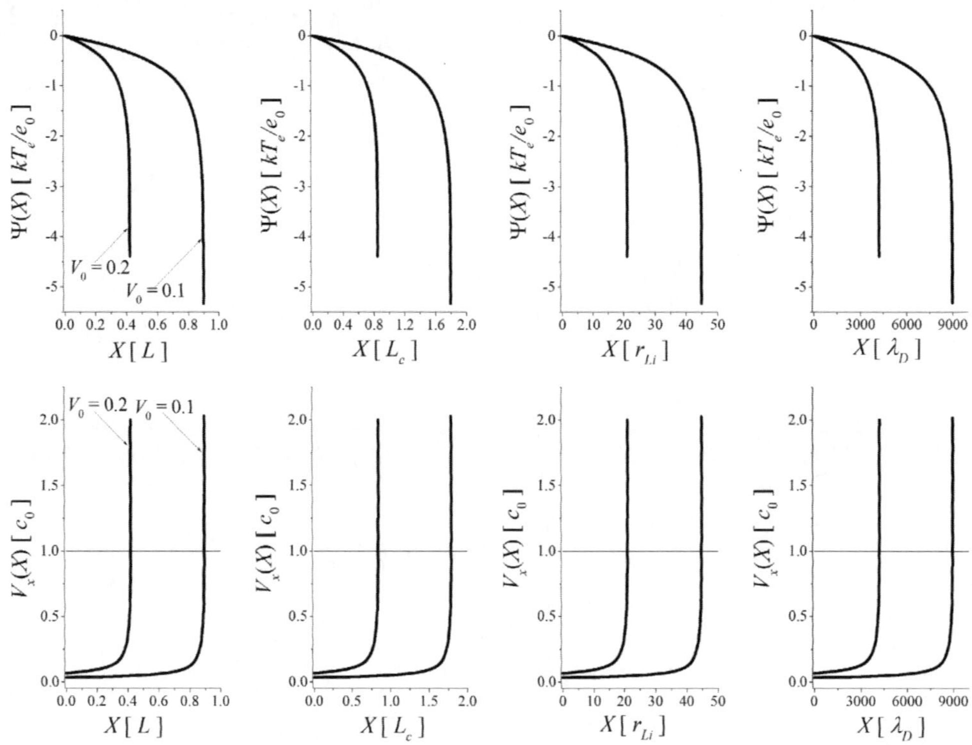

Figure 4. Potential $\Psi(X)$ (top) and velocity profiles $V_x(X)$ (bottom) found from the system (7) - (11) for the parameters $K = 50$, $Z = 2$, $\alpha = 20^{\circ}$, $\Theta = 0$ and $\varepsilon = 10^{-4}$. The boundary conditions are given by (15) with $V_0 = 0.1$ and $V_0 = 0.2$. The zero source term is used. The space coordinate is normalized to all 4 characteristic lengths.

When the zero source term is selected the solutions of the system (7) - (11) are much different than in the case when the other two source terms are used. One interesting property is illustrated in figure 5.

When the collision rate Z is low, the solutions of the system (7) - (11) exhibit coherent oscillations for certain boundary conditions and parameters. The following parameters are selected: Z = 0.001, $\varepsilon = 10^{-5}$, $K = 50$, $\alpha = 20^{\circ}$ and $\Theta = 0$.

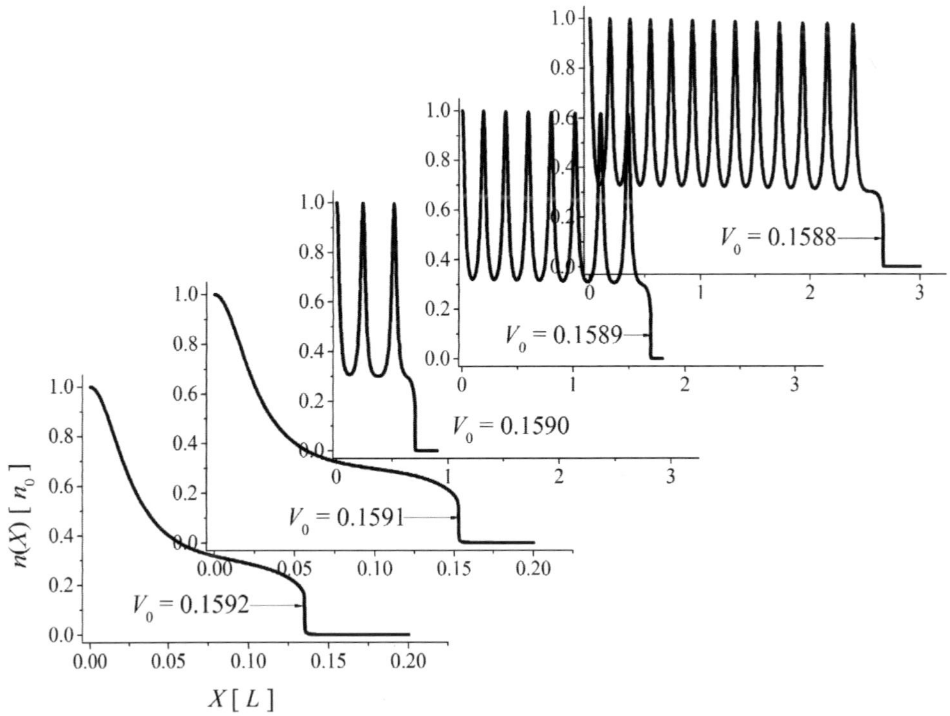

Figure 5. Oscillating solutions (ion density profiles $n(X)$) of the system (7) - (11) for the parameters: $Z = 0.001$, $\varepsilon = 10^{-5}$, $K = 50$, $\alpha = 20^{\circ}$ and $\Theta = 0$. The zero source term is used. The starting velocities are labelled in the figure.

The zero source term is used. In figure 5 the ion density profiles $n(X)$ are displayed. Whether these oscillations have some physical explanation, or are they a purely numerical effect is still an open question.

So far the ion temperature Θ has always been set to zero. If it is increased above zero the system (7) - (11) becomes singular. This stiffness of the system (7) - (11) can only be removed by increasing the velocity V_0 considerably. In figure 6 the minimum V_0 where the solution of the system (7) - (11) can be found is plotted versus the ion temperature Θ. The parameters are: $Z = 2$, $\varepsilon = 10^{-4}$, $K = 50$ and $\alpha = 20^{\circ}$. The exponential source is used. The results are very similar also for the constant and even for the zero source term. In the left plot both axis have linear scale. In the right plot the same results are shown, only both axis are in the logarithmic scale. The minimum V_0 is always slightly above the square root of the selected ion temperature. From (12) it can be seen easily that $\sqrt{\Theta}$ corresponds to the ion thermal velocity $\sqrt{kT_i/m_i}$.

Some examples of the ion density profiles $n(X)$ obtained with non-zero ion temperatures are shown in figure 7. The exponential source term is used and the following parameters are selected: Z = 2, $\varepsilon = 10^{-4}$, $K = 50$ and $\alpha = 20^{\circ}$. The ion temperatures 10^{-4}, 10^{-3}, 10^{-2}, 0.1 and 1 are selected. At each temperature the minimum velocity V_0 is used to solve the system (7) - (11). Because the increased ion temperature requires larger velocity V_0 the pre-sheath length decreases.

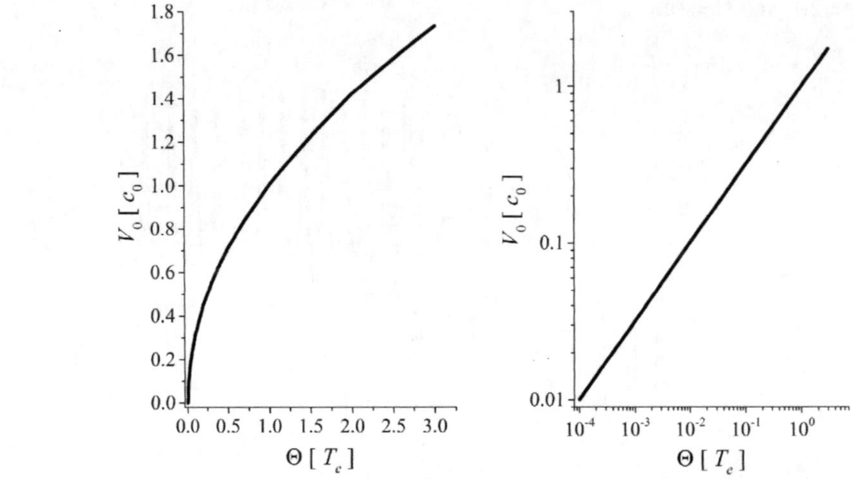

Figure 6. The minimum velocity V_0 for which the system (7) - (11) can be solved versus the ion temperature Θ. In the left graph the linear scale of the axis is used and in the right plot the logarithmic scale is applied.

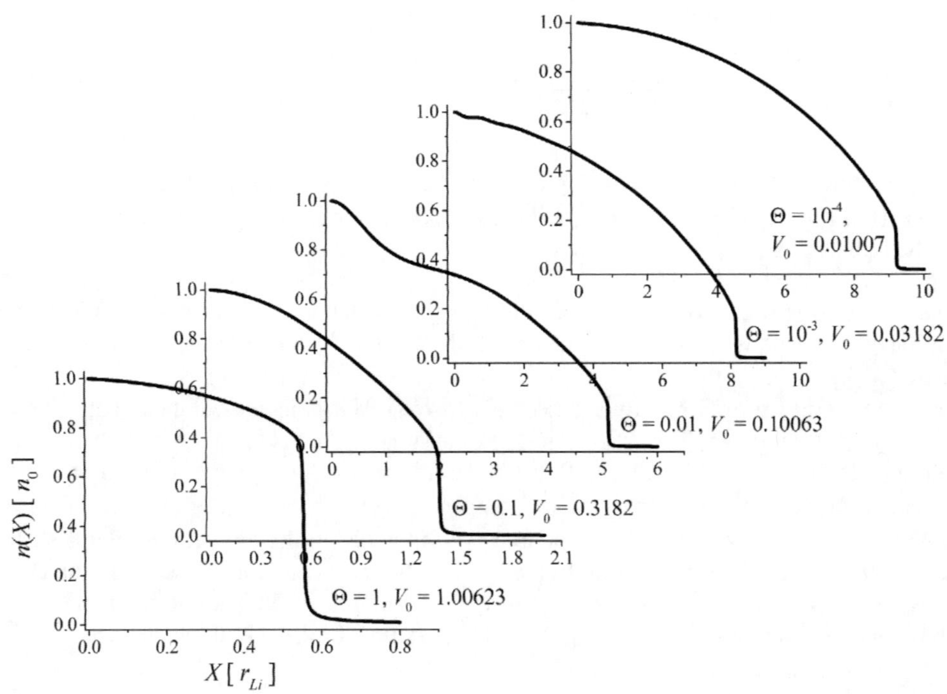

Figure 7. The ion density profiles $n(X)$ obtained from the system (7) - (11) for the parameters $Z = 2$, $\varepsilon = 10^{-4}$, $K = 50$ and $\alpha = 20^{\circ}$. Several ion temperatures Θ are selected and the minimum necessary V_0 is used at each Θ.

4. Conclusions

A one-dimensional fluid has been presented and used for the analysis of the potential formation in front of a negative planar electrode immersed in magnetized plasma. The magnetic field lines form an arbitrary angle with the electrode surface. The only exception is that they must not be completely parallel to the electrode surface. Three different source terms have been considered. For the constant source term the pre-sheath region is shorter than for the exponential source term if the other parameters are the same. If the zero source term is selected, the behaviour of the solutions is different. The starting ion velocity must be increased for several orders of magnitude in order to obtain solutions with comparable pre-sheath lengths as for the constant and the exponential source terms. When the zero source term is selected and the collision rate is decreased, the solutions exhibit strong oscillations. The nature of these oscillations remains to be investigated.

Four characteristic length scales appear in the problem – the ionization length L, the collision length or mean free path L_c, the ion Larmor radius r_{Li} and the Debye length λ_D. The space coordinate x can be normalized to any of them. In all four cases the solutions are exactly the same only the horizontal axis is expanded or stretched. The selection of the coordinate scaling is purely a matter of taste and has no physical or mathematical consequences. If one is interested in the pre-sheath region the ionization length L is appropriate choice for scaling. If one is more interested in the sheath region, the Debye length λ_D is more suitable. If the collision effects are studied, the mean free path L_c is most probably the best choice.

If the zero source term (4) is selected, the ionization time τ loses its physical meaning because it becomes infinitely large. But formally it is still possible to derive dimensionless equations (7) - (10) from (1) and (2) by multiplication with τ. In this case τ should be interpreted as a formally defined time constant, which is equal to $K/2\pi$ cyclotron periods T_{cyc} and to Z collision times T_{coll}. If the zero source term is selected, also the ionization length L loses its physical meaning. But one can still introduce a formal length scale L, which is equal to K Larmor radii r_{Li} and to Z mean free paths L_c and to $1/\varepsilon$ Debye lengths λ_D. The Debye length λ_D, the Larmor radius r_{Li} and the mean free path L_c are still well defined relatively to each other.

If the ion temperature is increased above zero, the system of equations becomes stiff. This stiffness is removed by increasing the starting ion velocity V_0. The minimum V_0, where the stiffness is removed, is slightly above the square root of the ion temperature. From (12) it can be seen that this corresponds to the ion thermal velocity.

Acknowledgements
This work has been partially supported by the grants P2-0073 and L7-4009 of the Slovenian Research Agency.

References

[1] Riemann K U 1994 *Contrib. Plasma. Phys.* **34** 127
[2] Riemann K U 1994 *Phys. Plasmas.* **3** 552
[3] Kovačič J, Gyergyek T, Čerček M 2009 *Eur. Phys. J. D* **54** 383
[4] Zimmermann T M G, Coppins M, Allen J E 2008 *Phys. Plasmas* **15** 072301
[5] Kuhn S, Riemann K U, Jelić N, Tskhakaya D D, Tskhakaya D Jr, Stanojević M 2006 *Phys.Plasmas* **13** 013503

Journal of Physics: Conference Series **768** (2016) 012006 doi:10.1088/1742-6596/768/1/012006

Mechanism of N_2 dissociation and kinetics of $N(^4S)$ atoms in pure nitrogen plasma

A V Volynets[1,2] , D V Lopaev[2] and N A Popov[2]

[1]Lomonosov Moscow State University, Faculty of Physics, Leninskie Gory, Moscow 119991, Russian Federation
[2]Lomonosov Moscow State University, Skobeltsyn Institute of Nuclear Physics, Leninskie gory, GSP-1, Moscow 119991, Russian Federation

E-mail: devorum89@gmail.com

Abstract. This work deals with kinetics of the ground state nitrogen atoms $N(^4S)$ and N_2 dissociation mechanism in pure nitrogen plasma. The experiment was carried out in positive column of DC glow discharge at a range of parameters p = 5 – 50 Torr, J = 20 – 100 mA. The use of axial homogeneous glow discharge allowed considering $N(^4S)$ balance for spatially uniform conditions controlled by only two terms: source (characterized by effective production rate k_{eff}^{diss}) and loss (characterized by effective loss time τ_{loss}^{N}). Analysis of these parameters gains considerably better understanding of N_2 dissociation mechanism in pure nitrogen plasma that was the main goal of the given work. So N/N_2 dissociation rate as function of discharge parameters was obtained using two independent emission optical methods: actinometry on Ar atoms and N_2 2+ band emission decay at discharge modulation. Measurements of N/N_2 radial profiles allowed estimating N atom surface loss probability γ_{loss}^{N} and correspondingly τ_{loss}^{N} . It was revealed that γ_{loss}^{N} depends on $N(^4S)$ concentration and thereby discharge conditions through the sorption balance of physisorbed N atoms. Simple phenomenological model taking into account basic surface processes provides γ_{loss}^{N} data in good agreement with experiment. Finally, k_{eff}^{diss} was obtained as function of reduced electric field E/N and it was shown that even EEDF self-consistently calculated with accounting for N_2 vibrational excitation is unable to provide observed values of k_{eff}^{diss} . Reasons of that fact are discussed in detail.

1. Introduction

At the increased pressure the associative ionization is the main ionization mechanism in nitrogen plasma [1]. Crucial point is that nitrogen molecules are able to accumulate considerable energy in their vibrations because of (v-v) exchange processes. Through the intermediate agents this energy is spent on excitation of electronic states of atoms and molecules which consequently participate in associative ionization processes. Despite of big interest to nitrogen plasma, mechanism of such energy transmission has not been studied yet in details. Metastable $N_2(A^3\Sigma_u^+)$ molecules and $N(^2P)$ atoms are believed to be most probable intermediate agents [1] that realize energy stored in N_2 vibrations into N_2 ionization. Since direct N_2^+ production in collision of two $N_2(A^3\Sigma_u^+)$ molecules is impossible, the mechanism of associative ionization has to include few steps with using other particles. Such particles can be nitrogen atoms. So, as a first step, it was studied $N(^4S)$ kinetics in pure N_2 plasma of positive column (PC) of DC - glow discharge at pressures 5 - 40 Torr. Such type of plasma is sufficiently suitable test object for this research because it is longitudinally homogeneous and radially symmetrical. Besides, this type of discharge is relatively easy to sustain.

Here $N(^4S)$ balance in plasma was considered in simple zero - dimensional form, controlled by only two terms: source and drain:

$$\frac{dN(^4S)}{dt} = k_{eff}^{diss} n_e N_2(X^1\Sigma_g^+, v) - \frac{N(^4S)}{\tau_{loss}} \tag{1.1}$$

where $N(^4S)$, $N_2(X^1\Sigma_g^+, v)$ are concentrations of ground state N atoms and ground state vibrationally excited N_2 molecules respectively, n_e – electron concentration. Source term comprises all possible ways of $N(^4S)$ production (i.e. N_2 dissociation reactions of all possible electronic and vibrational states) and characterized by effective production rate – k_{eff}^{diss}. Similarly, depletion term includes all mechanisms of atoms' loss and characterized by effective depletion time τ_{loss}^N.

Since in equilibrium conditions:

$$\frac{dN(^4S)}{dt} = 0 \tag{1.2}$$

So, according to (1.1) k_{eff}^{diss} and τ_{loss}^N reveal to be related.

Thus, goal of this work is to examine kinetics of $N(^4S)$ in pure nitrogen plasma. This comprises determination of $N(^4S)/N_2$ dissociation rate versus discharge parameters – electric field and pressure – over whole operating range. Secondly, experimental obtaining of τ_{loss}^N based on analysis of loss processes. Finally, using above mentioned relation with τ_{loss}^N, one can evaluate k_{eff}^{diss} as well as establish it's dependence on discharge parameters. Further analysis of obtained k_{eff}^{diss} values and comparison with theory afford considerably better understanding of N_2 dissociation processes in pure nitrogen plasma.

2. Experiment

2.1. Setup

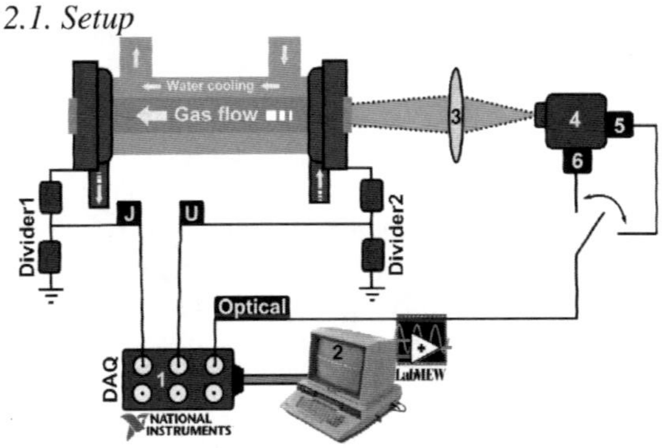

Figure 1. Experimental setup.

Experimental setup is shown on figure 1. Discharge operated in cylindrical quartz tube with MgF_2 windows. Tube's wall was cooled by flowing water. Electric current, applied voltage and optical signal were acquired using NI Data Acquisition (DAQ) board (1) and further transferred to PC (2). Optical emission from the tube's interior was collected by objective (3) and directed on the entrance slit of the monochromator (4). Distance from the tube's center to the monochromator varied from several cm to 2m approximately. Commonly tube was adjacent to monochromator when optical measurements in the tube's axis region were needed. Such measurements were accomplished with photomultiplier tube (PMT) (5). Large distance was chosen for optical measurements on the tube's cross-section with high spatial resolution. In this case (Intensified Charge Coupled Device) ICCD matrix (6) was used.

2.2. Gas temperature determination with emission optical method

Knowledge of gas temperature for every used discharge regime is necessary in this work. Spectrum of light emitted by plasma is sensitive to ohmic gas heating. Therefore spectrum's shape could contain information about T_{gas}. Indeed, the spectrum is defined by structure of N_2 electronic – vibrational – rotational levels and by distribution of molecular population of these levels (see, for example, [2]). Here we used several bands of second positive system $N_2(C^3\Pi_u, v' - B^3\Pi_g, v'')$. Molecular population of rotational levels of upper states obeys Boltzmann distribution which is characterized by rotational temperature T_{rot}. Because of intensive collisional exchange between energies of rotational and translational motions T_{rot} reveals to be related to T_{gas} and very close to it by magnitude.

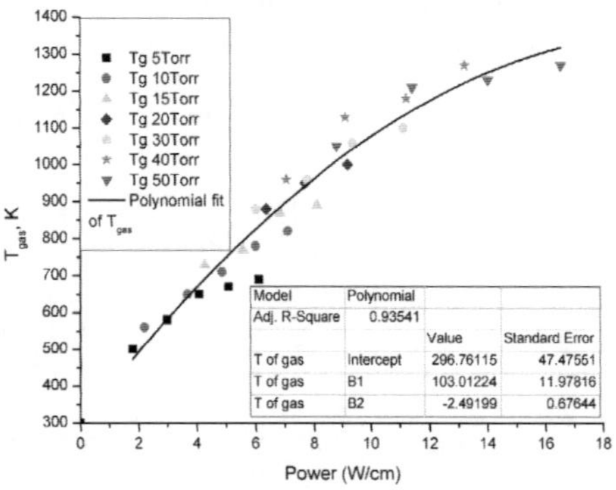

Figure 2. T_{gas} for all discharge regimes. Here power per unit length of the positive column (PC) is $E \cdot J$, where E – electric field in PC (see section *2.3.*), J – electric current.

Thus, it was created computer model of N_2 emission spectrum where T_{gas} was the main parameter. Comparing model spectrum with experimental one it was determined T_{gas} in the tube's axis region as function of discharge parameters (figure 2).

Emission spectrums of the same spectral range but obtained with lesser resolution was used for spatial determination of T_{gas} or getting it's dynamics in time. In that method [3] was used the fact that ratio of maximal intensities of P – and R – branches defines spectrum's shape and consequently T_{gas}. With mentioned above computer model of emission spectrum it was determined a calibration curve (T_{gas} versus the ratio). It is worth noting that shape of the curve strongly depends on a monochromator's apparatus function.

Therefore in experiments only the ratio of the intensities was determined over whole tube's cross section. Next, the obtained radial profile of the ratio was recalculated to T_{gas} radial profile (using the calibration curve). Evaluated in that way T_{gas} radial profiles revealed to be nearly homogenous almost up to tube's wall for all used regimes of discharge. In other words, gas temperature in the tube was accepted to be approximately constant having value of that in the tube's axis region.

2.3. Reduced field E/N

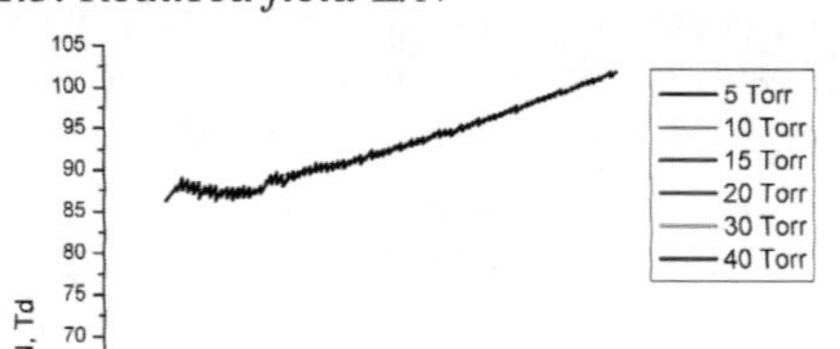

Figure 3. E/N in plasma for all discharge regimes.

Reduced field E/N (here E – electric field, N – concentration of neutrals) is a crucial parameter of any type plasma (which define electron energy distribution function (EEDF), reaction rates, etc). That is why this research requires specific values of E/N. Electric field in positive column was evaluated as $E = (V - V_{cpd})/L$ where V – applied voltage, V_{cpd} – cathode potential drop (values were taken from literature), L – length of the positive column (11 cm). Concentration of neutrals was calculated according to $P = Nk_{Bol}T_{gas}$ (here P, T_{gas} – gas pressure and temperature correspondingly, k_{Bol} – Boltzmann constant). Using T_{gas} data obtained previously, E/N was evaluated as function of discharge parameters (figure 3). Radial profile of the reduced field E/N in the tube was accepted to be uniform due to T_{gas} radial uniformity (mentioned in section 2.2).

2.4. N(4S) concentration and N/N_2 dissociation rate experimental obtaining

Absolute concentrations of $N(^4S)$ (as well as N/N_2 dissociation rate) on the tube's axis were determined using two independent techniques.

The first technique is actinometry with Ar atoms (5% Ar and 95% N_2). That was implemented in a similar manner with [4] where detailed description of the methodic was made. Obtained in that way results (figure 4) are highly dependent of corresponding actinometrical constants and, eventually, of calculated EEDF. (For N_2 plasma of used pressures the most essential way to obtain EEDF is to calculate it numerically using programs such as BOLSIG+). Thus inaccuracy of N/N_2 estimation with actinometry is defined mostly by the calculated EEDF.

These considerations gave motivation to perform this experiment once again in other way – using $N_2(C^3\Pi_u)$ radiation dynamics under electric current modulation. This method [5] was originally applied in studies of afterglows and has an advantage to be EEDF insensitive.

Obtained with two methods N/N_2 values are in good agreement (figure 5). However N/N_2 dependencies on discharge parameters (pressure, current) are somewhat different. Further analysis of that fact is needed. Comparison of the results allows us to test EEDF calculation adequacy as well as accuracy of commonly used actinometric constants.

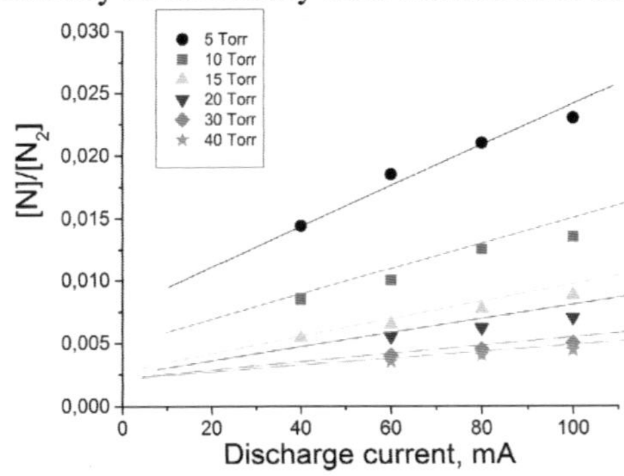

Figure 4. Obtained using actinometry method dissociation rate N/N_2 as function of discharge parameters (for the tube's axis region).

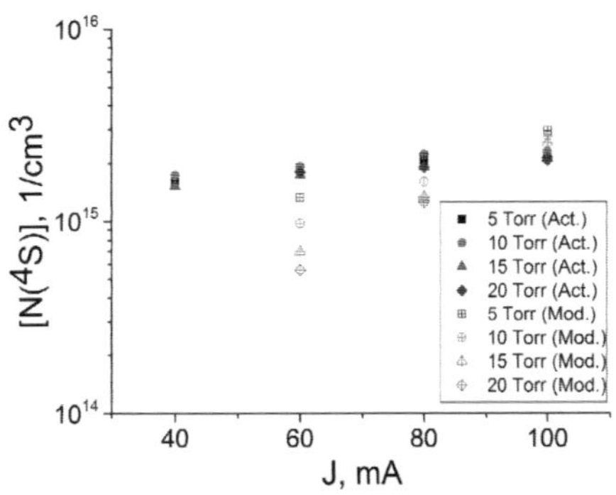

Figure 5. Absolute concentrations $N(^4S)$ (for the tube's axis region) determined with actinometry (solid symbols) and using "modulation technique" (crossed symbols).

2.5. N atom surface loss probability γ^N_{loss}

Simple estimations showed that for used pressures (p = 5 – 50 Torr) main N atom loss mechanism is recombination on tube's wall which proceeds according to $N + N + wall \longrightarrow N_2$. Quantitatively the probability of this process is characterized by value of γ^N_{loss}. This quantity is essentially related with mentioned in introduction τ^N_{loss} (N atom effective loss time):

$$\tau^N_{loss} \cong \frac{2R}{v^N_T} \cdot \frac{1}{\gamma^N_{loss}} \tag{2.5.1}$$

where $v^N_T \cong \sqrt{\frac{8}{\pi} \frac{k_B T}{M_N}}$ - average thermal speed of N atoms, R – tube's radius

Obviously, the first factor represents atom's diameter passage time. According to literature γ^N_{loss} has order about 0.01 – 0.001 what means that effectively atom passes the tube's diameter 100 – 1000 times before recombination. γ^N_{loss} was evaluated as function of discharge parameters using $N(^4S)/N_2$ radial profiles (analogically with [6]). It was revealed that γ^N_{loss} depends on $N(^4S)$ concentration and thereby discharge conditions through the sorption balance of physisorbed N atoms (figure 6).

To achieve better understanding, basic surface processes were considered in much the same way with work [6]. As a result simple phenomenological model was created. The model provides γ_{loss}^{N} (as function of discharge parameters) in good agreement with experiment indicating satisfactory level of understanding (figure 7).

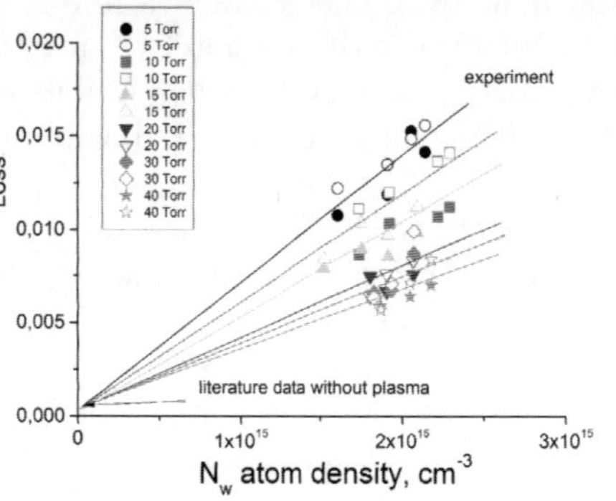

Figure 6. γ_{loss}^{N} as function of N(^{4}S) concentration. Symbols designate experimental values obtained using N/N_2 radial profiles. Solid lines represent results of the phenomenological simple model (see text for details).

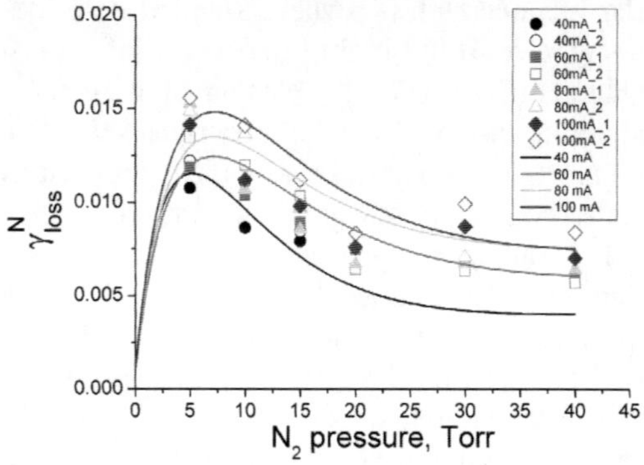

Figure 7. γ_{loss}^{N} as function of discharge parameters.

3. Effective N(^{4}S) production rate k_{eff}^{diss} and discussion.

According to equation (1.1) taking into account equilibrium condition (1.2) the effective N(^{4}S) production rate k_{eff}^{diss} could be expressed as:

$$k_{diss}^{eff} \cong \frac{\gamma_{loss}^{N}\dfrac{v_T^N}{2R}}{n_e}\cdot\frac{N(^4S)}{N_2} \tag{3.1}$$

n_e – electron concentration (evaluated from electric current $J = en_e v_{dreif}^{e}$), other quantities were mentioned above. Using above obtained data k_{eff}^{diss} was evaluated versus reduced field E/N and showed on figure 8 with symbols. Solid line represents theoretical rate dissociation constant in electronic collisions with vibrationally unexcited N_2 molecules of the ground state. It can be seen that nitrogen dissociation in N_2 plasma proceeds with higher rate than it is predicted by theory with the difference being notable (few orders) at low E/N.

k_{eff}^{diss} is highly sensitive to EEDF. For N_2 plasma of mentioned pressures the most essential way to obtain EEDF is to calculate it using programs such as BOLSIG+. EEDF used for theoretical k_{eff}^{diss} was calculated without consideration of N_2 vibrational excitation. However, transformation of EEDF due to N_2 vibrational excitation does not allow explaining solely the difference between theory and experiment. k_{eff}^{diss} calculated with such EEDF reveals to be still some orders of magnitude lesser than the experimental one. It means that particles having low energy threshold to N_2 dissociation fully

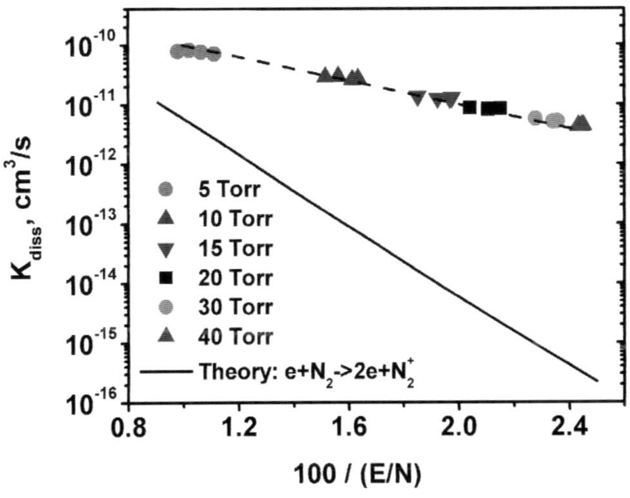

Figure 8. k_{eff}^{diss} experimental (symbols) and theoretical (solid line).

controls $N(^4S)$ production. $N_2(v)$ molecules on Treanor plateau are the most probable candidates for it. Rough estimation of effective energy threshold of N_2 dissociation in plasma gives value 2 – 3 eV, which corresponds to Treanor plateau in energy scale. It is also known, that intensive energy exchange exists between vibrationally excited molecules $N_2(v)$ of Treanor plateau and $N_2(A^3\Sigma_u^+)$ molecules. Role and significance of $N_2(A^3\Sigma_u^+)$ in N_2 dissociation haven't been completely understood yet. Detailed mechanism of N_2 dissociation in nitrogen plasma still has been needed in the following research.

Acknowledgements
This research is supported by SRC program Contract 2012-KJ-2280, by the Grants of Russian Foundation of Basic Research (12-02-00536-a) and by 14-02-31599 mol-a and by Optec grant № 28/2014/71-Msk

References

[1] Popov N A 2009 *Plasma Physics Reports* **35** № 5 p. 482-496

[2] Nassar H, Pellerin S, Musiol K, Martinie O, Pellerin N and Cormier J-M 2004 *J. Phys. D: Appl. Phys.* **37** 1904–1916

[3] Mi Lan, Xu Peng and Wang J Pei-Nan 2005 *J. Phys D: Appl. Phys.* **38** 3885-3888

[4] Levaton J, Ricard A, Henriques J, Silva H R T and Amorim J 2006 *J. Phys. D: Appl. Phys.* **39** 3285–3293

[5] Coitout H, Cemogora G and Magne L 1995 *J. Phys. III France* **V. 5.** P. 203-217

[6] Lopaev D V, Malykhin E M and Zyryanov S M 2011 *J. Phys. D: Appl. Phys.* **44** 015201

Journal of Physics: Conference Series **768** (2016) 012007 doi:10.1088/1742-6596/768/1/012007

Development of simulation tools for numerical investigation and computer-aided design (CAD) of gyrotrons

M Damyanova[1], S Sabchevski[1], I Zhelyazkov[2], E Vasileva[1], E Balabanova[1], P Dankov[2] and P Malinov[2]

[1]Acad. E. Djakov Institute of Electronics, Bulgarian Academy of Sciences, Association EURATOM-INRNE, 72 Tsarigradsko Chaussee, 1784 Sofia, Bulgaria
[2]Faculty of Physics, St. Kl. Ohridski University of Sofia, Association EURATOM-INRNE, 5 James Bourchier Blvd., 1164 Sofia, Bulgaria

E-mail: milena@ie.bas.bg

Abstract. As the most powerful CW sources of coherent radiation in the sub-terahertz to terahertz frequency range the gyrotrons have demonstrated a remarkable potential for numerous novel and prospective applications in the fundamental physical research and the technologies. Among them are powerful gyrotrons for electron cyclotron resonance heating (ECRH) and current drive (ECCD) of magnetically confined plasma in various reactors for controlled thermonuclear fusion (e.g., tokamaks and most notably ITER), high-frequency gyrotrons for sub-terahertz spectroscopy (for example NMR-DNP, XDMR, study of the hyperfine structure of positronium, etc.), gyrotrons for thermal processing and so on. Modelling and simulation are indispensable tools for numerical studies, computer-aided design (CAD) and optimization of such sophisticated vacuum tubes (fast-wave devices) operating on a physical principle known as electron cyclotron resonance maser (ECRM) instability. During the recent years, our research team has been involved in the development of physical models and problem-oriented software packages for numerical analysis and CAD of different gyrotrons in the framework of a broad international collaboration. In this paper we present the current status of our simulation tools (GYROSIM and GYREOSS packages) and illustrate their functionality by results of numerical experiments carried out recently. Finally, we provide an outlook on the envisaged further development of the computer codes and the computational modules belonging to these packages and specialized to different subsystems of the gyrotrons.

1. Introduction

Powerful gyrotrons with megawatt output power and frequencies ranging from 77 to 170 GHz are necessary for electron cyclotron resonance heating (ECRH), electron cyclotron current drive (ECCD) [1, 2] and ECH-assisted start-up of magnetically confined plasmas in various reactors for controlled thermonuclear fusion (tokamaks and stellarators) as well as for plasma control and stabilization (e.g., NTM suppression and MHD control). The gyrotrons are used also for plasma diagnostics based on a collective Thomson scattering (CTS).

It is commonly accepted and well understood that the capabilities of the simulation tools (physical models and numerical codes) is of crucial importance for a successful computer-aided design (CAD) and optimization of high performance gyrotrons with improved operational characteristics (higher efficiency, stability of the operation, mode purity, high-quality RF beam with a maximized Gaussian content, and so on). All this motivates the researchers involved in the development of gyrotrons for fusion to work on the improvement of the available simulation tools and on the development of novel more adequate physical models and software packages for numerical investigation and optimization of both the currently used and the future designs of these tubes. Our research team has been involved in the maintenance and further development of a great number of standalone computer programs and problem oriented software packages that are being used in the course of the CAD of gyrotrons for fusion. The work on these topics is being carried out in accordance with a recently formulated novel concept [3–6] for further improvements to the available simulation tools as well as for development of a new generation of numerical codes based on more adequate physical models, efficient numerical methods and algorithms, advanced (state-of-the-art) computational platforms.The main results obtained so far have been presented in the recent publications [7–13]. It should be noted also that in recent years the international collaboration of the Bulgarian research group has been extended to FIR FU Research Center at the University of Fukui (Japan). Although the gyrotrons that are being developed at FIR FU (except these for CTS plasma diagnostics) are dedicated to other applications, the fact that they operate on the same physical principles makes it possible to use common models and numerical codes for CAD and numerical studies of both classes of devices [13].

2. Current status of the simulation tools

GYROSIM [11] is a problem-oriented software package which includes numerical libraries and source codes of various computational modules (standalone programs, subroutines, pre-, post-processing, and visualization codes) for solving a variety of problems pertinent to the simulation and CAD of gyrotrons using a rich set of adequate physical models. Its structure is presented in Figure 1. The individual components of GYROSIM are used successively in an iterative design loop, during which numerical experiments are being performed for simulation of all main subsystems of the gyrotron tube, notably: (*i*) the electron-optical system (EOS), (*ii*) the magnetic system which includes the main magnet and an arrangement of additional coils, (*iii*) the electrodynamical system (resonant cavity), and (*iv*) the quasi-optical system for mode conversion and transmission of the radiation. The codes for numerical modelling of the EOS (GUN-MIG/CUSP) are based on a relativistic 2.5D physical model. GYROSIM includes also various cavity codes in which different physical models are implemented, e.g., time independent (static), time dependent, cold cavity, and self-consistent codes. The module GO&ART (which stands for Geometric Optics and Analytic Ray Tracing) of the GYROSIM package consists of several codes (RAYS, COMODES, and TRACE) for analysis of quasi-optical components (Vlasov and Denisov type launchers, reflectors and phase-correcting mirrors, and so on) as well as systems based on them (e.g., internal mode converters and transmission lines).

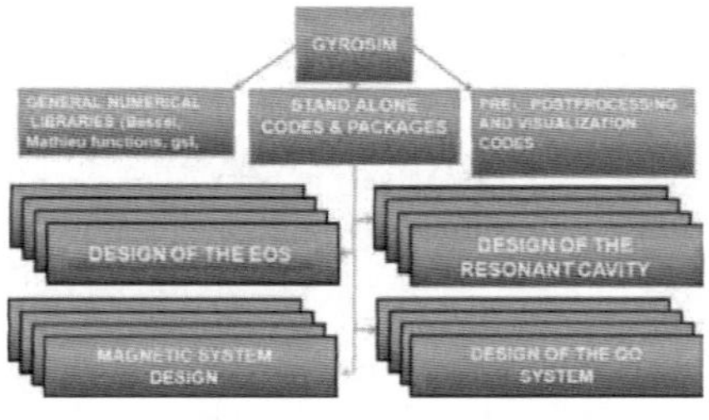

Figure 1. Structure of the package GYROSIM.

It should be mentioned that GYROSIM is a heterogeneous package since it includes components written in different languages (Fortran 77, Fortran 90, C, C++, SciLab, and so on), operational and/or portable to different computational platforms (ranging from laptops and workstations to mainframe

Journal of Physics: Conference Series **768** (2016) 012007 doi:10.1088/1742-6596/768/1/012007

and supercomputers), and executable under different (genuine as well as emulated/virtualized) operating systems (e.g., Unix, Linux, Windows, Cygwin). Another characteristic feature of the package is that it is being built following a concept of extensibility which allows us to add/replace easily different computational modules and in such a way to modify accordingly both the numerical algorithms and the physical models implemented in the programs.

GYREOSS has been initially conceived as a package of codes for simulation of EOS using a physical model formulated in three space dimensions (3D) in order to take into account the departure from axial symmetry due to various misalignments (for instance of the electrodes, of the magnetic coils, etc.) and non-uniformities [8, 10, 12]. Its first version has been implemented using the *gmsh* package for meshing, pre- and post-processing and *GetDP* as a solver [14]. In the recent years, however, GYREOSS has evolved as a test bench for experimenting with different numerical methods, solvers and algorithms in 3D aiming the final goal – a parallel 3D code for numerical simulation and CAD of EOS of gyrotrons. The latest version of GYREOSS is being developed using the FreeFEM++ problem solving environment [15]. Recently, a novel field solver in both 2D and 3D has been developed. It provides the components of the electromagnetic fields at the current particle locations for the relativistic particle pusher in which the Boris–Buneman scheme is implemented. The computational modules of the GYREOSS package developed so far are presented in Figure 2. They include practically all components that are needed for PIC simulations and trajectory analysis of EOS of gyrotrons. Among them are modules for generation of the mesh, magnetic and electric field solvers, relativistic particle pusher, etc. GYREOSS_MAGFI, which calculates the magnetic field produced by an arbitrary set of solenoids, is written in C language while all other modules are programed in the language of FreeFEM++. Depending on the specified parameters (in the modules GYREOSS-parameters and GYREOSS-init_part) different simulations can be carried out (e.g., 2D or 3D simulations, loading various initial particles distributions). The modular structure of GYREOSS allows an easy modification of its components and permits to include more physics in the envisaged (planned) future versions in order to take into account those physical factors that are neglected so far (for instance the dynamics of the trapped particles, space-charge compensation and so forth).

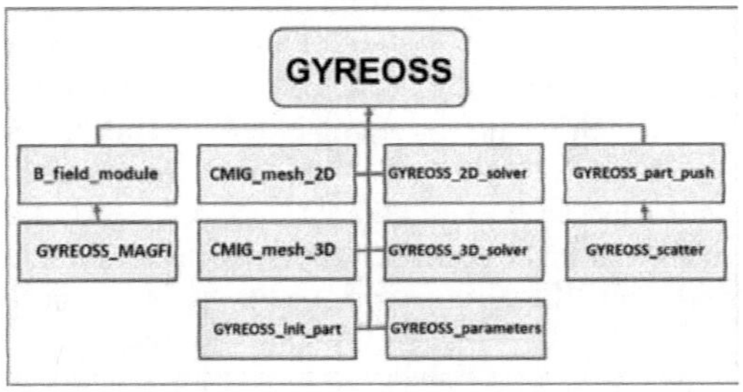

Figure 2. Computational modules of the GYREOSS package.

Some examples that characterize the visualization capabilities of the package GYREOSS are shown in figures 3 and 4, where screenshots from different stages of the numerical experiments are presented. The geometry input and generation of a tetrahedral mesh for simulations in 3D are illustrated in Figure 5, where the configuration of the EOS of a 170 GHz/2MW gyrotron with a coaxial cavity is shown.

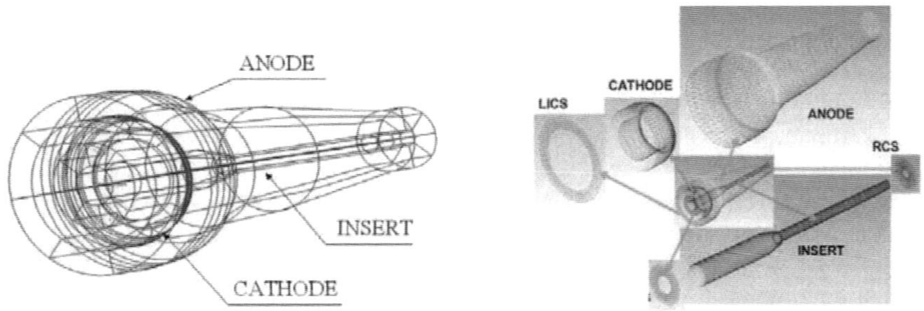

Figure 3. Geometry description (*left panel*) and boundary meshes (*right panel*) of the computational domain (including the closing surfaces on which Neumann boundary conditions are specified) used in the numerical experiments.

It should be noted that GYREOSS is able to use not only meshes generated by FreeFEM++ (which uses its own geometry description) but can import also mashes in *gmsh* format. While the CAD system *gmsh* provides means for an optimization of the mesh the FreeFEM++ package possesses powerful tools for mesh adaptation (based on a metric) with an automatic computation of the metric from a Hessian of the solution. Maps of both the potential distribution and the electric field calculated by the field solver are shown in Figure 4. The trajectories of the particles representing the electron beam in the EOS and traced using the novel relativistic particle pusher are plotted in Figure 5.

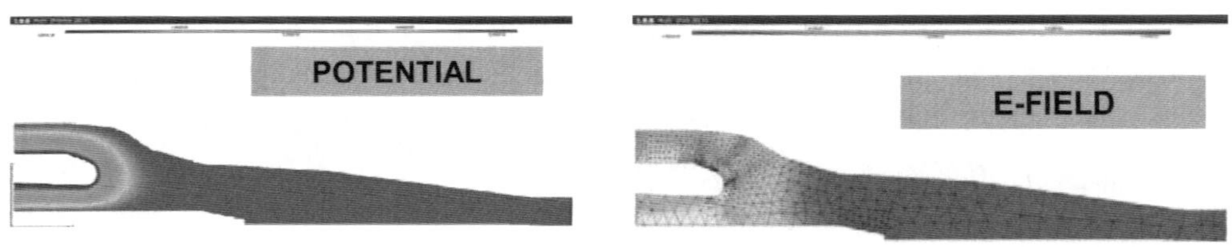

Figure 4. Maps of the potential distribution and the electrostatic field in a coaxial EOS of a 170 GHz, 2MW gyrotron.

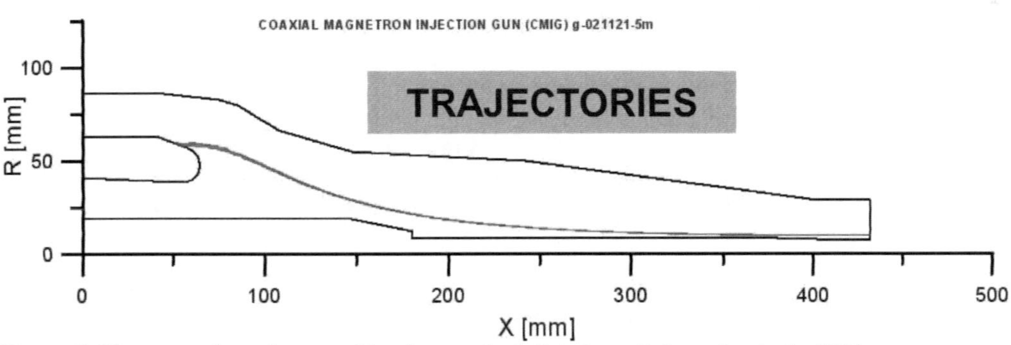

Figure 5. Electron trajectories traced by the novel relativistic particle pusher in the EOS.

An important step of any self-consistent PIC simulation is the space charge allocation (scattering) to the computational grid. Since this step of the algorithm is one of the most time-consuming its speed is of critical importance for the minimization of the required computational resources (CPU time and memory) as well as for the optimization of the overall throughput of the code. In view of the envisaged (planed) future parallelization of the code (which is considered as an indispensable one for the 3D model) another important requirement to the algorithm is to allow an easy transition from the currently developed sequential version to a parallel one. To solve these problems an efficient algorithm has been developed. It uses a fast searching algorithm and mesh connectivity information provided by FreeFEM++ and scatters the space charge to the nodes of the grid proportionally to the barycentric coordinates of the particles in a scheme which is compatible with the charge conservation law.

3. Conclusions and outlook

The problem-oriented software packages GYROSIM and GYREOSS, whose current status and functionality have been briefly presented and illustrated in this report, are under continuous and steady development, improvement, and adaptation to the ever-changing computational infrastructure (computer platforms, operating systems and underlying numerical libraries). Although outside the scope of this short paper, several other important topics that are being pursued in parallel with the code development should be mentioned as well. Among them is an ongoing theoretical work directed towards the formulation of novel and more adequate physical models and their subsequent implementation in the computer codes using efficient numerical methods and algorithms. The aim of these studies is to take into account physical factors and phenomena that are usually neglected in the current versions but are known to be of significant importance for the practical design and operation of the gyrotrons. Such are, for example, various electron-beam instabilities; dynamics of the reflected and trapped electrons; space-charge compensation; just to name a few. The computational modules (field solvers, particle pusher, etc.) developed so far and the modular structure of the packages form an appropriate basis for adding more physics in the envisaged and planned further extensions of our simulation tools for analysis and CAD of gyrotrons.

Acknowledgements

This work was carried out in the framework of Task 2.1.2 of the scientific program of the EURATOM-INRNE Association.

References

[1] Darbos C *et al* 2009 *Fusion Eng. Des.* **84** 651–655
[2] Omoria T *et al* 2011 *Fusion Eng. Des.* **86** 951–954
[3] Sabchevski S *et al* 2006 *J. Phys.: Conf. Series* **44** 96–101
[4] Sabchevski S *et al* 2006 *J. Phys.: Conf. Series* **44** 102–109
[5] Sabchevski S and Zhelyazkov I 2007 *J. Phys.: Conf. Series* **63** 012003
[6] Sabchevski S *et al* 2007 *Comput. Model. Eng. Sci.* **20** 203–220
[7] Sabchevski S *et al* 2008 *Wissenschaftliche Berichte FZK* 7409 pp 1–34
[8] Damyanova M *et al* 2010 *J. Phys.: Conf. Series* **207** 012032
[9] Damyanova M *et al* 2014 *J. Phys.: Conf. Series* **516** 012028
[10] Damyanova M *et al* 2012 *J. Phys.: Conf. Series* **356** 012044
[11] Damyanova M *et al* 5th Int. Workshop & Summer School on Plasma Physics (25–30 June, 2012, Kiten, Bulgaria)
[12] Damyanova M *et al* 2014 *J. Phys.: Conf. Series* **514** 012056
[13] Sabchevski S 2013 *International Symposium on Development of Terahertz Gyrotrons and Applications* (14–15 March, 2013, Fukui, Japan) visit http://fir.u-fukui.ac.jp/IntSymTHzGyrotrons2013.pdf
[14] Dular P and Geuzaine Ch 2013 GetDP: a General Environment for the Treatment of Discrete Problems (Version 2.4.1, July 16, 2013) visit http://geuz.org/getdp/
[15] Hecht F 2014 FreeFem++ Third Edition Version 3.30 (Paris: Laboratoire Jacques-Louis Lions, Université Pierre et Marie Curie) visit http://www.freefem.org/ff++/ftp/freefem++doc.pdf

Simulation of plasma filled hemispherical cavity as dielectric resonator antenna

G Trenchev, Zh Kissóvski

Faculty of Physics, Sofia University "St. Kliment Ohridski", 5 James Boucher Blvd, Sofia, Bulgaria

E-mail: gltrenchev@uni-sofia.bg, kissov@phys.uni-sofia.bg

Abstract. Plasma antennas are becoming an increasingly interesting research topic because of their uncommon characteristics. They are highly configurable, can be turned on and off rapidly, and exhibit lower thermal noise compared to metal antennas. In recent years, research has been conducted on cylindrical plasma columns sustained by DC, RF or microwave field, and their application as leaky wave antennas or as regular monopole antennas.

Dielectric resonator antennas (DRA) with high dielectric permittivity are known for their small size and excellent operating characteristics for modern mobile communications (WiMAX, LTE). Hemispherical dielectric resonator antennas are characterized by simple shape, high radiation efficiency and wide bandwidth. Hemispherical DRA with a low density weakly ionized plasma as dielectric material will combine the positive features of plasma and dielectric antennas, and is particularly interesting, as antennas of this type have not been studied yet. The hemispherical plasma antenna is simulated with Ansoft HFSS in the microwave S-band. Obtained radiation pattern and bandwidth show the advantages of hemispherical plasma antennas for future communication technology.

1. Introduction

Using a dielectric material as a resonator antenna has been studied extensively and has become more common in recent years [1]. Dielectric resonator antennas are characterized with high efficiency (typically 95-99%), small dimensions and low manufacture cost. They can be produced in almost any form and size, and can be excited by microstrip lines, slots and coaxial probes. Materials with dielectric permittivity of 2-100, as well as layered dielectrics have been studied theoretically and experimentally as resonator antennas with high success.

Low-temperature plasma has been studied and considered an adequate replacement for metal in antennas because of its low thermal noise, instant turn-on, and configurable radiation pattern. The applicability and efficiency in the RF range as leaky-wave antennas has been demonstrated in many experiments [3]. Operation of the cylindrical plasma antennas with plasma column sustained by surface waves as resonant antennas in the microwave S-band is still under research [4].

A hemispherical plasma resonator antenna with a low density weakly ionized plasma as dielectric material will combine the positive features of plasma and dielectric antennas. The main goal of our research is to study plasma as a substitute to dielectrics in a hemispherical DRA and to achieve a realistic model through computer-aided simulations.

2. Methods of simulation

The operation of the plasma resonator antenna depends on the relative permittivity of plasma. Plasma permittivity is evaluated as:

$$\varepsilon_p = 1 - \frac{\omega_p^2}{\omega(\omega - j\nu)} \quad , \tag{1}$$

where ω_p stands for plasma frequency, ω is the angular frequency of the microwave signal, and ν is the frequency of elastic collisions between electrons and neutrals. Formula for plasma frequency is:

$$\omega_p = \left(\frac{ne^2}{\varepsilon_0 m_e}\right)^{1/2} \quad , \tag{2}$$

where n is the plasma density, e – electron charge, m_e - electron mass, ε_0 - vacuum dielectric constant. Considering the relation between (1) and (2), it is clear that dielectric permittivity of plasma depends on electron density and the collision frequency.

In this research we assume that the hemisphere is filled with argon gas at low pressure of p = 3 Pa. Taking into account our experimental results for cylindrical plasma antenna [4], we estimate that a microwave signal with about 10 W of power could sustain plasma with electron density of $7.5*10^{17}$ m^{-3} inside the cavity. In this case, the complex permittivity of plasma will be $\varepsilon = -4 + j0.004$.

Hemispherical DRA offer analytical solution [2] for resonant frequencies and antenna radiation pattern. However, formulas for calculating resonant frequencies of hemispherical DRA are inapplicable for the plasma antenna because of the negative permittivity. For this reason, the plasma cavity is computer-simulated as a hemispherical resonator antenna with unknown properties. The software used is Ansoft HFSS, which allows simulation of materials with negative permittivity.

The plasma antenna design is described in figure 1. The exciting element consists of a microstrip line (1) fed by a lumped port (2). The dielectric substrate is sandwiched between the microstrip and the ground plane (3). Substrate thickness is 0.6 mm with ε_r = 3. The plasma antenna (4) with radius of 13 mm is placed above the circular slot (5) in the ground plane, and is excited by the loop element embedded in the microstrip (6). The loop element has inner radius of 5.0 mm. The radiation domain is a sphere with radius of ~5λ.

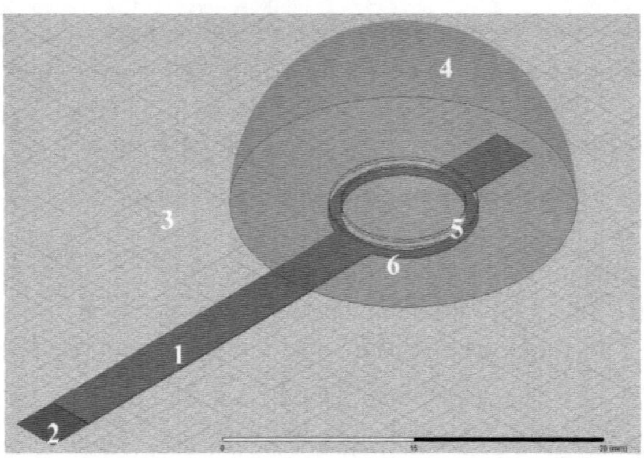

Figure 1. 3D model of hemispherical resonator plasma antenna.

During investigations, several methods for slot excitation were examined with results showing that circular slot above a loop element is the most efficient. The microstrip line length is designed for optimal VSWR.

3. Results

To better clarify the obtained results, the plasma resonator antenna is compared to a dielectric resonator antenna. A simulation is performed with a regular dielectric material with $\varepsilon_r = 9$ inside the hemispherical cavity. As expected, HEM_{11} mode (figure 2) is excited in the dielectric antenna at 3.5 GHz, radiating in vertical direction. In this case, results for electromagnetic field configuration and radiation pattern confirm known theoretical models on hemispherical DRA [1].

Substituting the dielectric with plasma with permittivity of $\varepsilon = -4 + j0.004$, the S_{11} parameter remains the same, but different distribution of electromagnetic fields inside the cavity is obtained. Results (figure 3) show that TM_{01} mode is excited with vertical and radial components of electric field while magnetic field has azimuthal dominant component. Radiation of the plasma antenna is similar to monopole antenna over ground plate. The resonant frequency is 3.5 GHz, which is a common band for the WiMAX and LTE standards in wireless communications. Operating frequency and antenna bandwidth depend on the value of negative permittivity (figure 4), and, consequently, on plasma density inside the cavity.

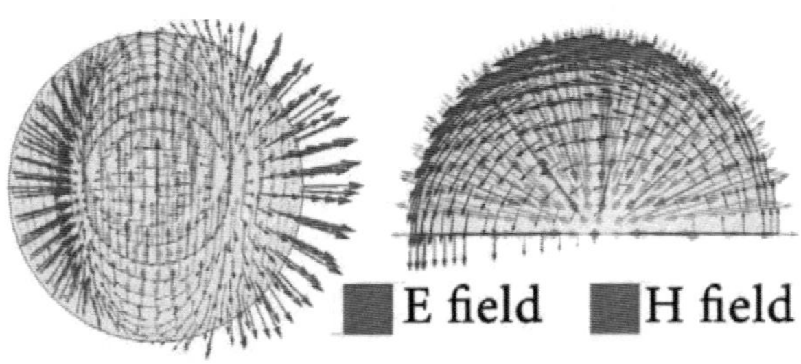

Figure 2. HEM_{11} mode at 3.5 GHz resonance frequency inside ordinary dielectric hemispherical resonator antenna.

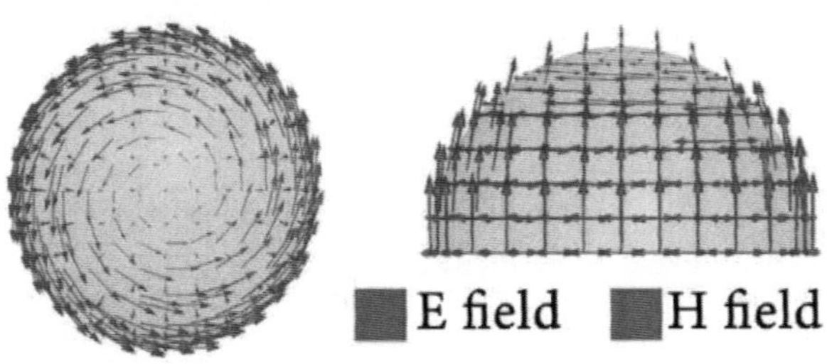

Figure 3. TM_{01} mode inside plasma hemispherical resonator antenna at 3.5 GHz resonance frequency.

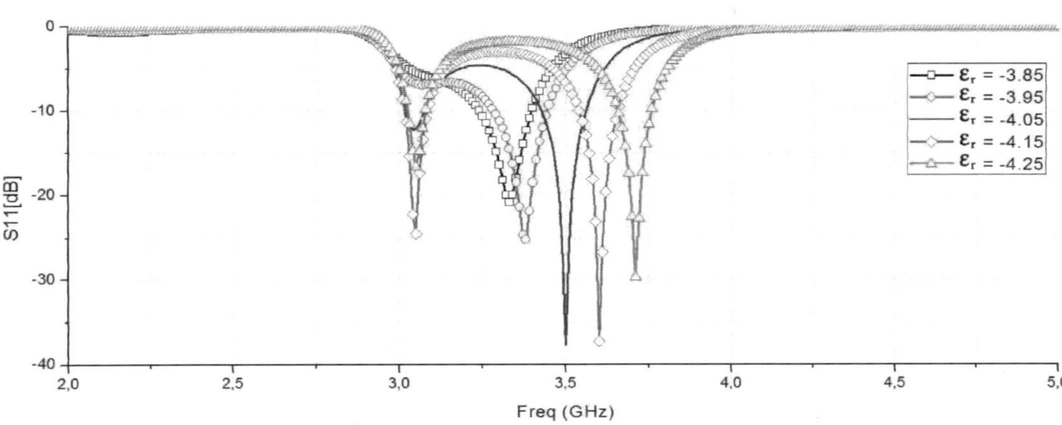

Figure 4. Dependence of S_{11} parameter on operating frequency of plasma antenna at various values of plasma permittivity. Operating mode is TM_{01}.

As shown in figure 4, small variance in permittivity of plasma results significant deviation of resonant frequency, which allows dynamic frequency tuning. The best result at 3.5 GHz is obtained with permittivity ε_r = -4.05 with VSWR of 1.02. The bandwidth of about 150 MHz is sufficient for WiMAX and LTE operation.

Radiation pattern of hemispherical plasma antenna (figure 5) shows relatively high gain of 5 dBi and omnidirectional profile. Wave attenuation in the plasma resonator antenna is very low, which leads to high efficiency, close to 100%, like in dielectric resonator antennas. The radiation pattern shape of hemispherical plasma antenna is similar to the radiation pattern of a wire monopole antenna over finite (with radius ~5λ) and over infinite ground plane.

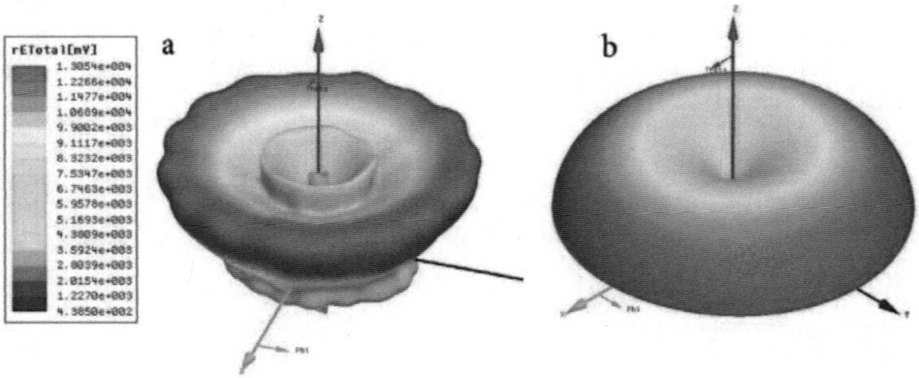

Figure 5. Plasma antenna 3D radiation model over finite (a) and over infinite ground plane (b).

4. Conclusion

Plasma resonator antenna with hemispherical shape is investigated with an EM simulator. Obtained radiation pattern is similar to radiation pattern of a monopole antenna. Plasma antenna is characterized by high efficiency and configurability. Plasma density can be used to dynamically tune resonant frequency and antenna bandwidth. Plasma density needs high precision control to achieve optimal operating characteristics.

The plasma resonator antenna is a possible candidate for demanding 5G modern communications. It is compact, efficient, with low noise, and quick response, and within the reach of current technology. Compared to equivalent DRA, plasma antennas possess the unique properties as instant turn-off and frequency tuning. Further development of the research is aimed towards experimental verification of the simulation results.

Acknowledgements

This work is financially supported by Science Fund of Sofia University "St. Kliment Ohridski", project N114/2014.

References

[1] Leung K W and Luk K M 2004 *Dielectric Resonator Antennas* (Baldock, Hertfortshire, England: Research Studies Press)

[2] Mongia R K and Bhartia P, Dielectric Resonator Antennas – A Review and General Design Relations for Resonant Frequency and Bandwidth, *Int. J. Microwave and Millimeter-Wave Comp.-Aid. Eng.*, Vol. **4** (1994)

[3] Anderson T 2011 *Plasma Antennas* (Boston, London: Artech House)

[4] Kissóvski Zh, Vachkov V, Iordanova S and Koleva I 2012 Microwave discharge in a finite length vessel, *J. Phys. Conf. Series* **356** 012009.

Journal of Physics: Conference Series **768** (2016) 012009 doi:10.1088/1742-6596/768/1/012009

Study of transient spark discharge focused at NOx generation for biomedical applications

M Janda, V Martišovitš, K Hensel and Z Machala

Division of Environmental Physics, Faculty of Mathematics, Physics and Informatics Comenius University, Mlynská dolina F2, Bratislava 84248, Slovakia

E-mail: janda@fmph.uniba.sk

Abstract. The paper is focused at nitrogen oxides generation by transient spark (TS) in atmospheric pressure air. The TS is a DC-driven self-pulsing discharge with short duration (~10-100 ns) high current pulses (>1A), with the repetition frequency 1-10 kHz. Thanks to the short spark duration, highly reactive non-equilibrium plasma is generated, producing ~300 ppm of NO_x per input energy density 100 $J.l^{-1}$. Further optimization of NO/NO_2 production to improve the biomedical/antimicrobial effects is possible by modifying the electric circuit generating the TS.

1. Introduction

Non-thermal plasmas generated by electrical discharges can be maintained at low gas temperatures, but have high electron energies responsible for their chemical activity. For this reason, several plasma based technologies were successfully implemented in the recent years, in processes such as surface modification, plasma assisted combustion, or hazardous waste processing [1-3]. In the past few years, a fast development of biological and medical cold plasmas applications has been observed [4], since cold plasmas provide multiple agents that can efficiently kill bacteria, yeasts, and other hazardous microbes. It is very important to assess the role of the plasma agents involved. At present, the major role in atmospheric pressure plasmas generated in air is typically attributed to reactive oxygen and nitrogen species (RONS) [5, 6], as well as to charged particles [7, 8].

The aqueous RONS are formed from dissolved NO, NO_2 and OH radicals generated by plasma in the gas phase [9]. The first step in the water bio-decontamination optimization process is thus better understanding of the NO and NO_2 generation in the gas phase. Partially, this problem has already been addressed by many authors in exhaust gas cleaning applications by electrical discharges [10-13]. However, they mostly studied NO reduction, or oxidation to NO_2, and finally to HNO_3. A recent paper [14] investigated the NO and NO_2 formation and their bactericidal effects in hybrid glow-spark discharges in air. Here we present a study focused on the synthesis of NO, NO_2 and their precursors (O and N species) by the transient spark (TS) discharge in air.

The TS is a dc-driven self-pulsing discharge with the typical repetition frequency 1-10 kHz. The TS is initiated by a streamer transforming to a short spark current pulse. Streamers are considered to be crucial for the efficiency of plasma induced chemistry at atmospheric pressure, since the electric field in the streamer's head can reach more than 200 kV/cm [15, 16], so that the chemical and ionization processes are very efficient there. On the other hand, transition of streamers to classical spark or arc discharge [17, 18], generating thermal plasma, is usually not desired. The thermal plasma is certainly not suitable for treatment of tissues or sterilization of surface sensitive to heat. Moreover, much more power is required to sustain these discharges. High energy costs allow their utilization only for applications where one can either expect valuable products (e.g. H_2) or needs to completely destroy dangerous pollutants [19-21].

However, the TS spark current pulses are sufficiently short (~10-100 ns) to avoid the plasma thermalization. Plasma generated during the spark phase of the TS is therefore highly reactive, since the electron density as high as 10^{17} cm^{-3} can be achieved [22]. The transient spark chemical activity is therefore comparable with the nanosecond repetitive pulsed (NRP) discharge [23]. The TS advantage is no need of special and expensive high voltage pulsers with high repetitive frequency and nanosecond rise-times used to generate NRP discharges [24-26]. The TS has already been successfully tested for several biological and environmental applications [6, 9, 27]. Basic research of TS using several electrical and optical diagnostic methods was also performed [22, 28-32]. However, further research is needed for better understanding of chemical processes initiated by the TS. In this work we present additional research of TS using optical emission spectroscopy (OES) combined with the post-discharge gas composition analysis.

2. Experimental set-up

A dc HV power supply connected to the anode via a series resistor (R = 9.84 MΩ) was used to generate a positive TS discharge. The distance d between the stainless steel electrodes (point-to-point configuration with anode at the top) was 6 mm. The electrodes were made of a 2-mm diameter rod. The anode tip was sharpened whereas the cathode was blunt. The curvature radius of the anode tip was of the order of 100 μm. The curvature radius of the cathode was of the order of a millimeter.

The electrodes were placed in a closed 100 mm long glass tube with the inner radius 4 cm and two valves for the gas inlet and outlet. The one side of the glass tube was enclosed by Teflon window with openings for the electrodes. The other side was enclosed by the CaF_2 window enabling optical diagnostic. The experiments were carried out in synthetic air from the pressure tanks (N_2:O_2 = 4:1, N_2: 99.99% purity, O_2: 99.5% purity), with flow rate 1.3 - 2.6 l.min^{-1}. We used a gas analyzer Kane KM9106 Quintox to measure the gas composition after passing through the discharge reactor. It can detect NO and NO_2 in the range of 0-1000 ppm with an accuracy of 3 %.

The discharge voltage was measured by a HV probe Tektronix P6015A and the discharge current was measured on a 50 Ω or 1 Ω resistor shunt. The 1 Ω resistor shunt was used when we measured the spark current pulse, whereas the 50 Ω resistor shunt was used to measure the streamer current. Electric signals were recorded by a 200 MHz digitizing oscilloscope Tektronix TDS2024.

For fast recording of time-integrated broadband spectra (200 - 1100 nm), we used a two-channel compact emission spectrometer Ocean Optics SD2000 (spectral resolution 0.6 - 1.7 nm). The light collection was provided by a pair of parabolic mirrors focused on a small area near the anode tip. These mirrors, together with the optical fiber core diameter, determined the spatial resolution to be 300 μm. The experimental set-up is depicted in figure 1.

3. Results and discussion

Transient spark is initiated by a streamer, when the potential on the stressed electrode V reaches voltage V_{TS}, characteristic for the TS. The streamer forms a relatively conductive plasma bridge between the electrodes. This leads to partial discharging of the electric circuit internal capacity C. The current from the partial discharging of C heats up the gas inside the plasma channel generated by the streamer. The transition to spark occurs when gas temperature inside the plasma channel increases to ~1000 K [30].

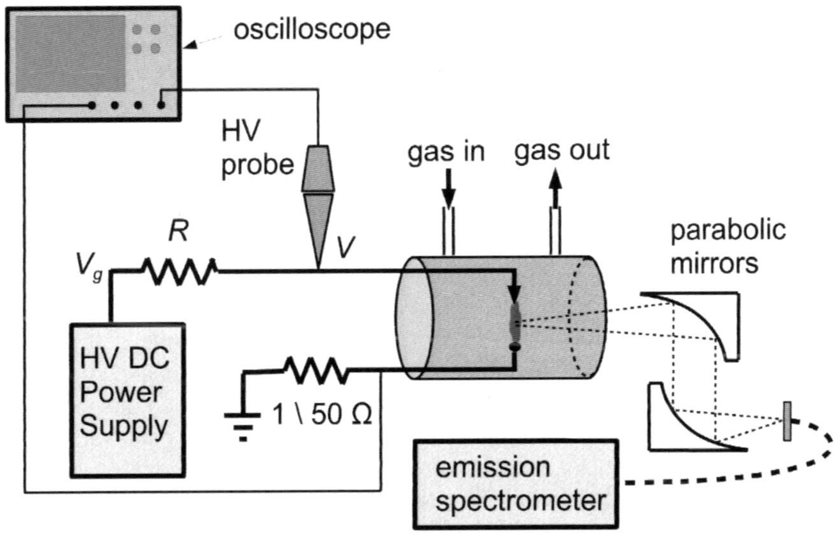

Figure 1. Simplified schematic of the experimental set-up.

During the spark phase, the electric circuit internal capacity discharges completely and the voltage on the HV electrode drops to almost zero (figure 2). The discharge current reaches a high value (~1-15 A). Transition to an arc after the spark phase is restricted by the ballast resistor R. It limits the current delivered to the plasma after C is discharged. As a result, the plasma starts to decay after the spark pulse. Eventually, the plasma resistance exceeds the R and the potential on the stressed electrode gradually increases, as the capacitor C recharges. A new TS pulse, initiated by a new streamer, occurs when the potential on the anode reaches the breakdown voltage V_{TS} again. The TS is thus based on repetitive charging and discharging of C. The repetition frequency f of this process can be controlled by the generator voltage V_g and is typically in the range 1-10 kHz [29].

The TS repetition frequency increase is accompanied by the increase of the mean discharge current I_{mean} and discharge power P (figure 3). However, several other TS discharge characteristics change as well. The breakdown voltage decreases, and the spark pulses are getting broader and smaller [29, 30].

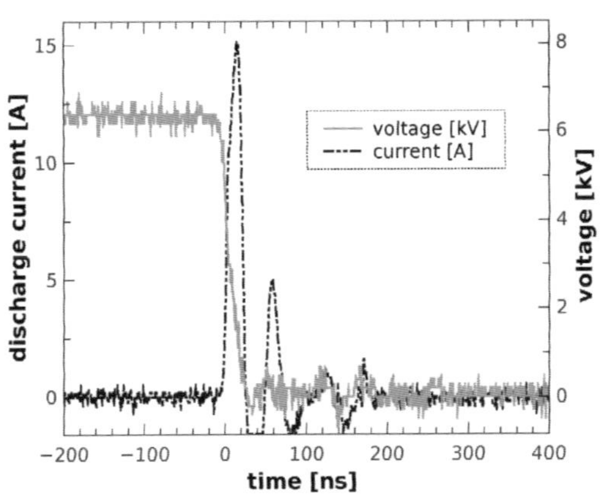

Figure 2. Typical transient spark voltage and current waveforms, $f \approx 2$ kHz.

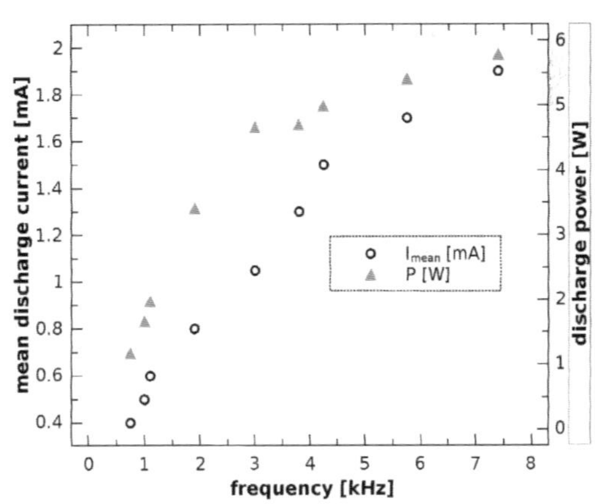

Figure 3. The discharge mean current I_{mean} and discharge power P as functions of TS repetition frequency.

The analysis of the electric circuit representing the TS described in [29], and the measurement of the discharge channel diameter by the time-resolved imaging [33] enabled us to estimate the electron density n_e in TS from the current and voltage waveforms [22, 29]. After the streamer, $n_e \approx 10^{14}$ cm^{-3} was obtained, which is in an agreement with values found in the literature [15, 34-36]. During the spark phase, the electron density as high as 10^{17} cm^{-3} can be achieved at $f < 4$ kHz. Generated plasma is thus strongly ionized for a short time. However, the TS spark current pulses are sufficiently short to avoid thermalization of generated plasma. With increasing frequency, smaller and broader spark current pulses occur and the peak electron density decreases to ~10^{16} cm^{-3} [22].

The time-integrated emission spectroscopy study confirmed that TS pulses generate highly reactive non-equilibrium plasma with excited atomic radicals (O^*, N^*), excited molecules N_2^* and ions N_2^{+*} (figure 4). The rotational temperature T_r derived from the N_2 2nd positive system time-integrated spectra, ranging from 500 to 1500 K, was much lower than the vibrational temperature T_v (3800-5000 K) [29]. The TS emission characteristics change with increasing frequency as well. Below ~3 kHz, the atomic lines emission (O^*, N^*) and N_2 2nd positive system dominates in the spectra, but at higher frequencies, these atomic lines almost disappear. This might be interpreted as a decrease of the TS ability to produce radicals and to induce chemical changes at higher frequencies, but it was not confirmed by the measurement of the nitrogen oxides densities.

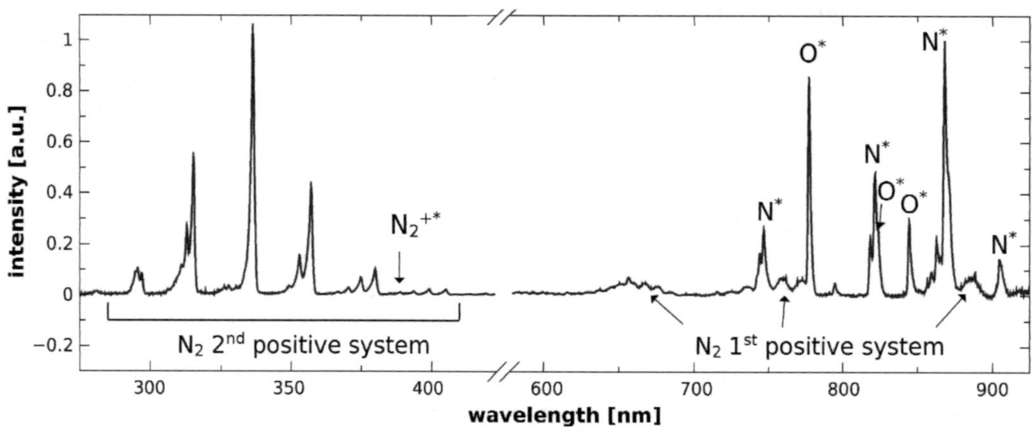

Figure 4. Typical time-integrated spectrum of the TS discharge in air, f ~ 2 kHz.

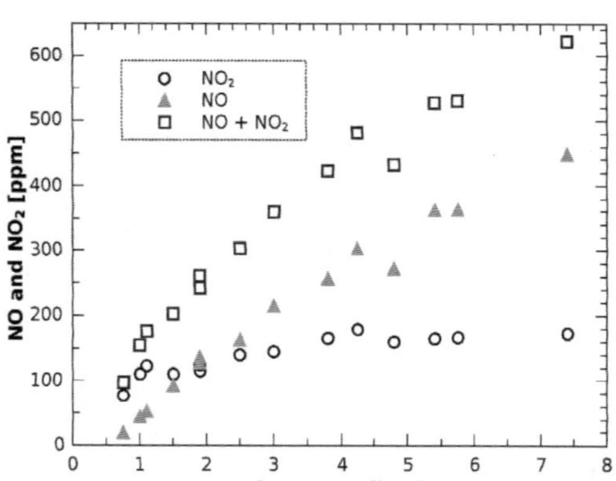

Figure 5. The NO and NO_2 densities produced by TS as functions of f, gas flow rate 1.7 l.min^{-1}.

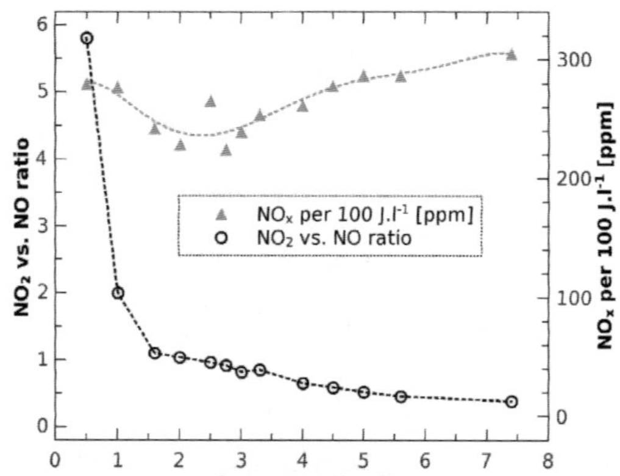

Figure 6. The NO and NO_2 produced by TS per energy input 100 J.l^{-1}, averaged for flow rates 1.3-2.6 l.min^{-1}.

Due to increasing input power, the densities of both NO and NO_2 increase with *f* (figure 5). In order to asses the NO_x generation efficiency, it is necessary to take into account the discharge power changes with increasing TS repetition frequency. Based, on the recalculated densities of NO and NO_2 generated per input energy density 100 $J.l^{-1}$ (figure 6), the NO generation efficiency tends to improve with increasing *f*, while the NO_2 generation efficiency decreases. However, the sum of NO and NO_2 densities do not change significantly. These results show that the ability of TS to induce chemical changes cannot be evaluated using the total NO_x density (NO + NO_2).

More likely, it is the NO_x synthesis mechanism that changes with the increasing TS repetition frequency. The decrease of NO_2/NO ratio (figure 6) with increasing TS frequency could mean that there is less O to oxidize NO further to NO_2. Thus, it could be related with the disappearance of atomic emission lines at higher TS frequencies. However, further research including kinetic modeling is required to explain the mechanism of the NO_x generation by the TS discharge and dependence of NO_2/NO ratio on TS repetition frequency.

4. Conclusions

The aim of our study was to investigate the NO and NO_2 generation, which are significant biomedical and biocidal agents, by the Transient Spark (TS) discharge in atmospheric pressure air. The TS is a streamer-to-spark transition type discharge initiated by a streamer, with the spark pulse duration and amplitude limited by a small amount of energy stored in the circuit internal capacity *C*. The TS repetition frequency (1-10 kHz) can be controlled by the applied voltage. Similar type of plasma, even at higher pulse repetition frequencies, can be generated using nanosecond repetitive pulsed high voltage power supplies. The advantage of our approach is simpler and cheaper electric setup using DC high voltage power supply.

The TS discharge can be maintained at low energy conditions (~1 mJ/pulse) by an appropriate choice of the resistances and capacities in the electrical circuit, so that the generated plasma stays out of thermal equilibrium. The non-equilibrium state of plasma generated by TS was confirmed by the optical emission spectroscopic study. Our research confirmed high reactivity of plasma generated by the TS. In dry synthetic air, the sum of NO and NO_2 densities more than 600 ppm was achieved with power input below 6 W (~300 ppm per 100 J of energy deposited per 1 liter of gas). In future, we plan to optimize the NO_x production by varying various electric circuit parameters (external resistor, internal capacitor, distance of electrodes).

The TS characteristics change with the increasing TS repetition frequency. At higher frequencies, the decrease of the breakdown voltage, smaller and broader spark pulses, and lower electron density during the TS spark phase were observed. These changes also influence the generation of excited atomic species and nitrogen oxides. The atomic lines emission intensity decreases and the NO/NO_2 ratio increases with increasing TS repetition frequency. This indicates changes in the TS chemical reactivity or the mechanism of NO_x generation changes with increasing TS repetition frequency. Further research including kinetic modeling is required to resolve this problem.

Acknowledgments

Effort sponsored by the Slovak Research and Development Agency APVV-0134-12, and Slovak grant agency VEGA 1/0918/15.

References

[1] Cernak M, Rahel J, Kovacik D, Simor M, Brablec A and Slavicek P 2004 *Contrib. Plasma Phys.* **44** 492
[2] Sahni M and Locke B R 2006 *J. Hazard. Mater.* **137** 1025
[3] Messerle V E, Karpenko E I and Ustimenko A B 2014 *Fuel* **126** 294
[4] Machala Z, Hensel K and Akishev Y 2012 *Plasma for bio-decontamination, medicine and food security* (*NATO Science for Peace and Security Series A* - Chemistry and Biology) (Dordrecht: Springer)

[5] Graves D B 2012 *J. Phys. D: Appl. Phys.* **45** 263001
[6] Machala Z, Chládeková L, Pelach M 2010 *J. Phys. D: Appl. Phys.* **43** 222001
[7] Fridman G, Brooks A D, Balasubramanian M, Fridman A, Gutsol A, Vasilets V N, Ayan H and Friedman G 2007 *Plasma Process. Polym.* **4** 370
[8] Sysolyatina E, Mukhachev A, Yurova M, Grushin M, Karalnik V, Petryakov A, Trushkin N, Ermolaeva S and Akishev Y 2014 *Plasma Process. Polym.* **11** 315
[9] Machala Z, Tarabová B, Hensel K, Špetlíková E, Šikurová L and Lukeš P 2013 *Plasma Process. Polym.* **10** 649
[10] Penetrante B M, Hsiao M C, Merritt B T, Vogtlin G E, Wallman P H, Kuthi A, Burkhart C P and Bayless J R 1995 *Appl. Phys. Lett.* **67** 3096
[11] Dors M, Mizeraczyk J, Czech T and Rea M 1998 *J. Electrostat.* **45** 25
[12] Herron J T 2001 *Plasma Chem. Plasma Proc.* **21** 581
[13] Zhao G-B, Hu X, Yeung M-C, Plumb O A and Radosz M 2004 *Ind. Eng. Chem. Res.* **43** 2315
[14] Pavlovich M J, Ono T, Galleher C, Curtis B, Clark D S, Machala Z and Graves D B 2014 *J. Phys. D: Appl. Phys.* **47** 505202
[15] Morrow R and Lowke J J 1997 *J. Phys. D: Appl. Phys.* **30** 617
[16] Kulikovsky A A 1998 *IEEE Trans. Plasma Sci.* **26** 1339
[17] Loeb L B 1965 *Electrical Coronas, Their Basic Physical Mechanisms* (Berkeley, CA: University of California Press)
[18] Raether H 1964 *Electron Avalanches and Breakdown in Gases* (London: Butterworths)
[19] Kobayashi A, Osaki K and Yamabe C 2002 *Vacuum* **65** 475
[20] Bromberg L, Cohn D R, Rabinovich A and Alexeev N 1999 *Int. J. Hydrogen Energy* **24** 1131
[21] Kanzawa A 1993 *Plasma Sources Sci. Technol.* **2** 58
[22] Janda M, Martišovitš V, Hensel K, Dvonč L and Machala Z 2014 *Plasma Sources Sci. Technol.* **23** 065016
[23] Janda M, Machala Z, Lacoste D A, Stancu G D and Laux C O 2012 Stabilization of a lean methane-air flame using transient spark discharge *International Symposium on High Pressure Low Temperature Plasma Chemistry HAKONE XIII* (Kazimierz Dolny, Poland, 9-14 September) ed T Janowski et al pp 185-89
[24] Walsh J L, Shi J J and Kong M G 2006 *Appl. Phys. Lett.* **89** 161505
[25] Pancheshnyi S V, Lacoste D A, Bourdon A and Laux C O 2006 *IEEE Trans. Plasma Sci.* **34** 2478
[26] Pai D, Lacoste D A and Laux C O 2008 *IEEE Trans. Plasma Sci.* **36** 974
[27] Machala Z, Morvová M, Marode E and Morva I 2000 *J. Phys. D: Appl. Phys.* **33** 3198
[28] Machala Z, Janda M, Hensel K, Jedlovský I, Leštinská L, Foltin V, Martišovitš, and Morvová M 2007 J. Mol. Spectrosc. **243** 194
[29] Janda M, Martišovitš V and Machala Z 2011 *Plasma Sources Sci. Technol.* **20** 035015
[30] Janda M, Machala Z, Niklová A and Martišovitš V 2012 *Plasma Sources Sci. Technol.* **21** 045006
[31] Gerling T, Hoder T, Brandenburg R, Bussiahn R and Weltmann K-D 2013 *J. Phys. D: Appl. Phys.* **46** 145205
[32] Gerling T, Hoder T, Bussiahn R, Brandenburg R and Weltmann K-D 2013 *Plasma Sources Sci. Technol.* **22** 065012
[33] Janda M and Machala Z 2011 *IEEE Trans. Plasma Sci.* **39** 2246
[34] Bourdon A, Bonaventura Z and Celestin S 2010 *Plasma Sources Sci. Technol.* **19** 034012
[35] Kacem S, Eichwald O, Ducasse O, Renon N, Yousfi M and Charrada K 2012 *J. Comput. Phys.* **231** 251
[36] Barni R, Esena P and Riccardi C 2005 *J. Appl. Phys.* **97** 073301

Journal of Physics: Conference Series **768** (2016) 012010 doi:10.1088/1742-6596/768/1/012010

Diagnostic of plasma streams from ion thrusters for space propulsion using emissive probes

L. Conde[1], S.P. Tierno[1], J.L. Domenech-Garret[1], J.M. Donoso[1], M. A. Castillo [2], I. Eíriz [3] and I. Sáez de Ocáriz [3]

[1] Dept. de Física Aplicada, ETSIAE, Univ. Politécnica de Madrid, 28040, Madrid, Spain
[2] Aernnova Engineering Solutions, 28030 Madrid, Spain
[3] Centro de Tecnologías Aeronáuticas, Miñano, Vitoria, Spain

E-mail: luis.conde@upm.es

Abstract.
The emissive probes are employed for the determination of the local plasma potential of plasma streams produced by ion thrusters. Its operation basically relies on electron collection and emission and are less sensitive to the ion motion than collecting probes. The diagnostic using emissive probes is reviewed with emphasis in low density plasmas. Our results support the conclusion that potential structures around the probe, as virtual cathodes, would be responsible for the operation of emissive probes in low density plasmas.

1. Introduction

The next generation of satellites will make use of the recent developments in space propulsion using high energy plasma streams. The advantages of electric propulsion (EP) for long term missions rely on combined reasons of economy and practical interest. These systems produce ion streams with exhaust velocities of tens of kilometers per second, much higher than conventional propulsion systems using reactive chemicals as propellants. This fact leads to high values of the specific impulse that results in relevant weight savings over traditional propulsion systems [1]. In addition, the EP engines are powered by electric the solar panels and are propelled by chemically inert gases (usually Xenon) of easy stowage. Therefore, the EP offers considerable advantages for in orbit station keeping of geostationary satellites, deep space missions or planetary probes [1].

These propulsive systems need of long term endurance experiments and intensive laboratory testing in order to determine their performances, as well as their actual levels of thrust. In particular, the mapping of the electric potential spatial profile of the low density plasma stream (also denominated *plasma plume*) [1] is of the paramount importance. The electric fields driving the plasma stream collimate the dispersion of the plasma exhaust stream and therefore modulate the effective thrust imparted by the ion engine. Additionally, large angle dispersion of the plasma plume would affect the operation of nearby satellite payloads by the undesired bombardment of ions from the plasma thruster exhaust [1].

The electric probes allow pointwise measurements of the plasma parameters and are well suited for this space mapping [2]. The collecting Langmuir (CLP) and emissive (EMP) probes are currently used for the determination of the plasma density and electric potential spatial profiles. They are made of plasma exposed metallic conductors that collect ions and electrons for different electric bias potentials. The plasma parameters are later evaluated from these voltage current (IV) characteristic curves of CLP and EMPs [2]. These calculations assume isotropic energy distributions for ions and electrons and involve physical models for the charge collection processes from the plasma.

However, in the plasma streams of ion thrusters the ions have important drift velocities with respect to the probe at rest and these conditions are not always fulfilled. The physical size of the probes also introduces plasma wakes downstream that are not considered in most CLP current collection models. Consequently, most classical models for CLPs are less applicable in plasma flows except in the limit of low relative ion drift velocities.

On the contrary, emissive probes are less sensitive to the ion motion because its operation basically relies on the electron collection and emission processes [2, 3, 4]. As the Fig. (1) shows the probe emits a thermionic electron current which is retained for bias voltages V_p over the plasma potential $V_p > V_{sp}$ or emitted to the plasma otherwise. The local value of the plasma potential is determined by the abrupt change in the probe current produced by the transition between the electron collection and emission. The contribution of ions in low density plasmas could be usually neglected. This local measurement makes EMPs suitable for the plasma potential mapping of plasma streams [2, 3, 4].

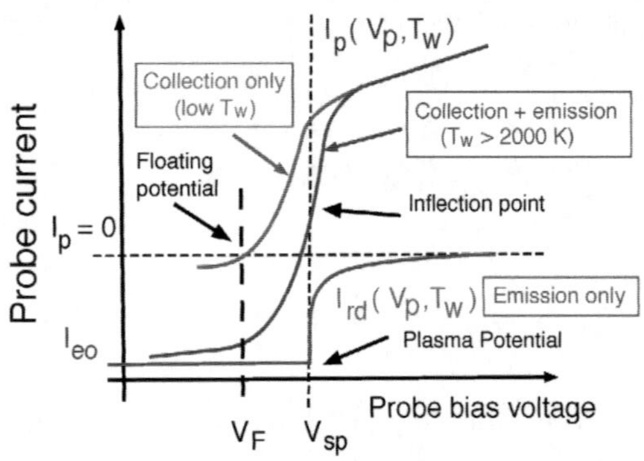

Figure 1. Scheme of two idealized IV curves $I_p\,(V_p, T_w)$ of an emissive probe and the thermionic emission current $I_{rd}\,(V_p, T_w)$ as a function of the probe bias potential.

Nevertheless, despite the success of this plasma diagnostic its theoretical support still remains incomplete [4, 5, 6]. In the operation of EMPs the plasma results inevitably perturbed by the electron thermionic emission and we will discuss its extent for the case of low density plasmas. In this case the measurements using the denominated *inflection point* method [3, 4], corresponding to low emission thermionic currents are difficult because of the poor signal to noise ratio as the Fig. (2) shows. On the opposite limit, the larger thermionic currents involved in the so called *floating point* method [5, 7, 8] might result in the strong perturbation of the nearby plasma.

However, the experimental data in this *strong emission* regime of EMP suggest the development of plasma potential structures around the probe [9, 12, 13, 14] that could be responsible for the partial trapping of the emitted electrons [5]. We conjecture that such electron confinement would limit the plasma perturbation extent and explains the reliable measurements of V_{sp} using the floating potential of EMPs.

2. The operation of the emissive Langmuir probes

The original concept of the EMP could be traced back to Irving Langmuir and the Refs. [4, 7] are comprehensive reviews of this experimental technique. The EMPs are essentially intended to measure the local plasma potential and also the electron temperature according to some authors [15]. They are usually made up of a electrically conductive thermionic electron emitter heated up to high temperatures, such as a loop of a thin tungsten wire [5, 6] or a LaB_6 crystal [8]. This latter is exposed to the plasma and for the temperature T_w emits a thermionic electron current density is given by Richardson-Dushman expression,

$$J_{rd}(T_w) = C\,T_w^2 \exp\left(-\frac{e\,W_f}{k_B T_w}\right) \tag{1}$$

where C is a constant and W_f the work function of the material [16]. In our particular case [5], the probes are made of a tungsten wire with diameter $d = 0.8$ mm heated up by a DC current. The wire temperature T_w is related to the heating current I_H through the following phenomenological expression,

$$T_w = T_o + P_1\left(\frac{I_H}{d^{3/2}}\right)^{P_2} \tag{2}$$

Here d is the diameter of the wire in centimeters, $P_1 = 29.50$ K, $P_2 = 0.59$ are constants and $T_o = 204.35$ K [17, 18].

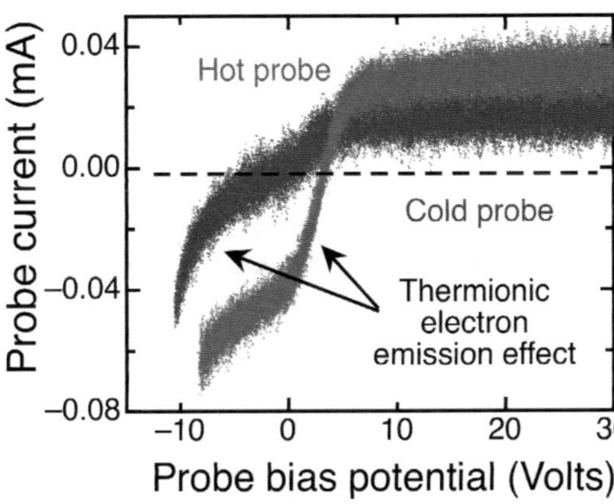

Figure 2. The current voltage characteristic curves of an emissive probe for two temperatures T_w in the low thermionic emission mode.

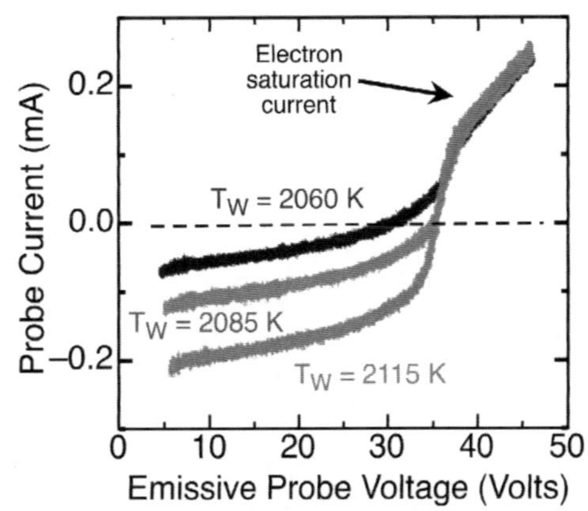

Figure 3. The current voltage characteristic curves of an emissive probe of in the large thermionic emission mode.

In the scheme of Fig. (1) the vertical dotted line denotes the value of the plasma potential V_{sp} and the horizontal dashed line indicates the current $I_p(V_F, T_w) = 0$ corresponding to the floating potential $V_F(T_w)$. The red curve for the thermionic electron emission current has a steplike characteristic where the maximum negative emitted electron current $I_{eo}(T_w)$ takes place for $V_p \ll V_{sp}$ because all thermionic electrons are rejected towards the plasma. On the contrary, the emission current $I_{rd}(V_p, T_w) \simeq 0$ for $V_p \gg V_{sp}$ when the electrons cannot leave the probe surface.

Two idealized IV curves $I_p(V_p, T_w)$ are also represented in Fig. (1) to illustrate the EMP operation principle. For low probe temperatures T_w the thermionic electron current $I_{eo}(T_w) \simeq 0$ (green curve) could be neglected and this cold EMP collects a current from the plasma $I_p(V_p, T_w) \simeq I_p(V_p)$ essentially as a CLP. When $V_p < V_F$ the probe drains the small ion saturation current and $I_p(V_p, T_w)$ increases for $V_p > V_F$, and for bias voltages over the knee corresponding to V_{sp} the probe current saturates.

The blue IV characteristic curve $I_p(V_p, T_w)$ for a hot EMP in Fig. (1) results from the combined effects of the thermionic electron emission and charge collection from the plasma. The electron emission current $I_{rd}(V_p, T_w)$ introduces a sharp jump located close to the plasma potential and the magnitude of this steep change depends of the probe temperature T_w through the Eq. (1).

The typical IV curves of two different emissive probes operating in low and large thermionic emission modes are represented in Figs. (2) and (3) whereas the details of these experiments are discussed in Ref. [5]. For cold probe temperatures the small surface of the emissive probe collects low currents that give rise to a poor signal to noise ratio. The effect of the electron thermionic emission is also observed in Fig. (2) by the small shift between the two curves when the probe temperature is slightly increased.

In order to obtain the characteristic curve of Fig. (3) the probe temperature needs to be substantially raised over the values of Fig. (2) and $I_p(V_p, T_w)$ increments by more than an order of magnitude. The probe temperature dependent step produced by the emission current $I_{eo}(T_w)$ of Fig. (1) that gives rise to the difference between Figs. (2) and (3) also depends of additional factors, such as the length of the wire or the cleanliness of the metal surface. Our probes need to be typically heated up to $T_w > 2000$ K to obtain appreciable electron emission currents as those of Fig. (3).

The arrows in the scheme of Fig. (1) indicate relevant points such as those related to the *inflection point* and the *floating potential* methods that are currently employed to determine V_{sp} from the IV characteristics curve of EMPs. As the probe temperature increases, the growing emission currents $I_{rd}(V_p, T_w)$ added to the cold probe characteristic produce steeper $I_p(V_p, T_w)$ curves close to the plasma potential. This is observed in the curves of Fig. (3) where the temperature dependent floating potential $V_F(T_w)$ approaches the plasma potential and $V_F(T_w) \simeq V_{sp}$ when the probe is hot enough. Additionally, the emission current introduces a change in the slope of $I_p(V_p, T_w)$ that would be observed in the curves of Figs. (2) and (3). The corresponding *inflection point* also moves close to $V_F(T_w)$ as the probe temperature increases.

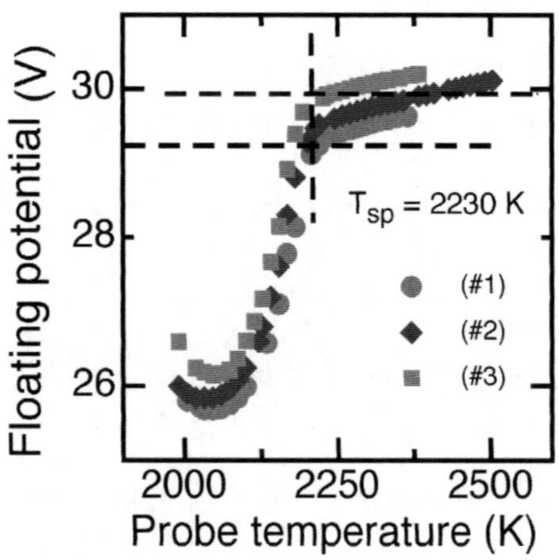

Figure 4. Three sets of measurements of the floating potential of the emissive probe as a function of the probe temperature T_w.

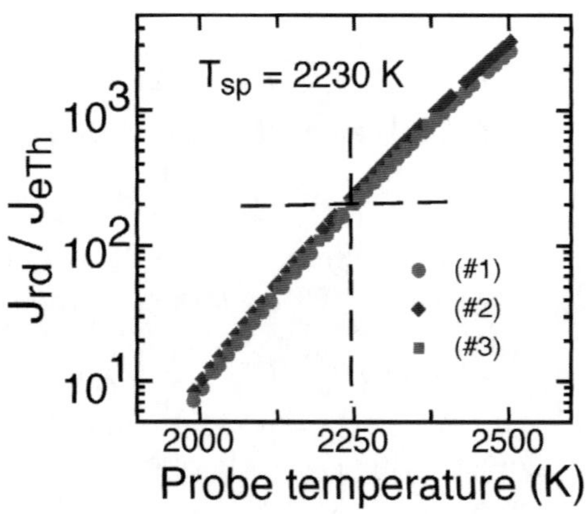

Figure 5. The ratio between the thermionic $J_{rd}(T_w)$ and the plasma thermal electron J_{eTh} current densities for the experimental data of Fig. 4.

The *inflection point* method [3, 4] requires a low thermionic electron current which minimizes the perturbation introduced in the plasma by the EMP operation. The value of probe current $I_i(T_w) = I_p(V_i, T_w)$ corresponding to the inflection point bias voltage $V_i(T_w)$ is found by derivation of the IV curves of Figs. (2) and (3) with respect to V_p [3, 4]. The maximum corresponds to $V_i(T_w)$ and these values are represented against the ratios $I_i(T_w)/I_{eTh}$ where I_{eTh} is the plasma electron thermal current. The plasma potential V_{sp} is calculated by experimental data fitting in the limit of low thermionic electron emission [3, 4].

This identification of such inflection point with the plasma potential is essentially qualitative but constitutes a good approximation for V_{sp} [3] as suggests the scheme of Fig. (1). The main drawback of this method in low density plasmas relies in the derivation of the $I_p(V_p, T_w)$ of noisy IV curves, as those of Figs. (2) and (3) with respect to the bias potential V_p.

Alternatively, the plasma potential could be also determined by the *floating point* method by the saturation observed in Fig. (4) when representing $V_F(T_w)$ against T_w (or equivalently, the heating current I_H) [5, 6, 7, 8]. For probe bias potentials $V_p \leq V_{sp}$ in a Maxwellian plasma the electron current density to the emissive probe is [8],

$$J_p(V_p, T_w, T_e) = J_{ce}(V_p, T_e) - J_{rd}(T_w) - J_{ci}(V_p) \tag{3}$$

where $J_{ci}(V_p)$ is the current density of attracted ions and $J_{rd}(T_w)$ given by Eq. (1) corresponds to the population of repelled thermionic electrons. The contribution $J_{ce}(V_p, T_e)$ of the Maxwellian plasma electrons is,

$$J_{ce}(V_p, T_e) = J_{eTh} \exp\left(\frac{e\,(V_p - V_{sp})}{k_B T_e}\right) \tag{4}$$

and when the probe is biased to the floating potential $J_p(V_F, T_w, T_e) = 0$ and we obtain,

$$V_F - V_{sp} = \frac{k_B T_e}{e} \ln\left(\frac{J_{rd}(T_w) + J_{ci}}{J_{eTh}}\right) \tag{5}$$

Therefore, as the probe temperature T_w increases the thermionic electron emission grows and when $J_{rd}(T_w) \simeq J_{eTh}$ we obtain $V_F \simeq V_{sp}$. The small contribution of the attracted ion current $J_{ci} \ll J_{eTh}$ is usually neglected [5, 7, 8] in low density plasmas.

The Eq. (5) is derived under the assumption of a Maxwellian energy distribution for the plasma electrons. More complex distributions such as bi-Maxwellian have been observed in a number of situations such as plasma double layers [10], divertor plasmas in tokamaks [11], etc. In these situations, the plasma parameters could then be obtained by using new methods that have been recently reported [11].

3. Discussion

The measurements of the floating potential as a function of the temperature T_w are easy to implement. The Eq. (2) allows us to estimate the temperature of the emissive probe that constitutes the key physical parameter to evaluate the thermionic electron emission current density of Eq. (1). This permits to asses both the Eq. (5) and the magnitude of the plasma perturbation by the EMP electron emission.

This latter is small for low T_w (typically below 2100 K) and in this regime the Fig. (4) shows that the emissive probe basically acts as a CLP, where the floating potential remains independent of the probe temperature. The values of $V_F(T_w)$ increases with T_w when the thermionic electron emission grows in Fig. (4) as well as in Figs. (2) and (3). Finally, the curve saturates in Fig. (4) for the temperature T_{sp} where $V_F(T_{sp}) \simeq V_{sp}$. This basic response of the EMP has been reported by a number of authors and provides values of the plasma potential with reasonable accuracy in different plasmas [5, 6, 7, 8].

In Fig. (5) are represented the ratios $R = J_{rd}(T_w)/J_{eTh}(T_e)$ between the emitted thermionic current density of Eq. (1) and the electron thermal random current of the plasma $J_{eTh}(T_e)$ for the experimental data of Fig. (4). Because of the low plasma densities ($n_e \sim 0.1 - 8 \times 10^8$ cm^{-3}) and electron temperatures ($T_e \sim$ 1-2 eV) the electron emission currents are between one and two orders of magnitude higher than $J_{eTh}(T_e)$ within this probe temperature range [5]. Therefore the excessive probe heating results in large thermionic electron emission currents that perturb the surrounding plasma. Nevertheless, the cross check of the EMP against the measurements with collecting probes indicates that the knee of Fig. (5) corresponds to the actual value of the plasma potential [5, 19].

Additionally, the Eq. (5) also would provide the electron temperature by the experimental data fitting of $V_F(T_w)$ as the Fig. (6) shows. The electron temperatures calculated from the slopes m of the experimental data of Figs.(4) and (5) are usually give higher values than those obtained by other methods.

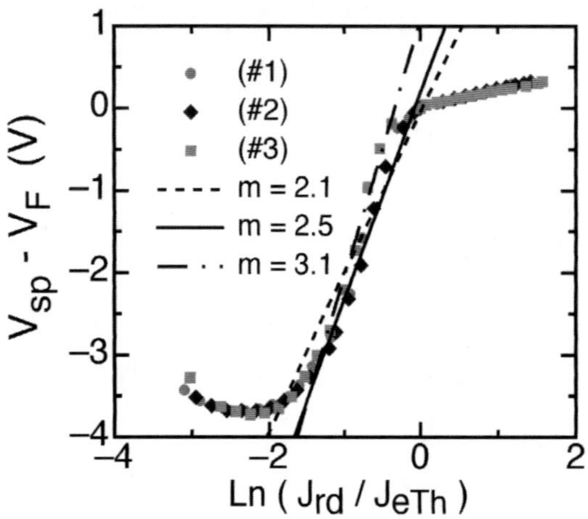

Figure 6. The fitting of the experimental data experimental data of Fig. (4) to Eq. (5) to evaluate the electron temperature using their slopes m.

4. Conclusions

The previous experimental evidences point out that the simple model for $V_p \leq V_{sp}$ that gives rise to Eq. (5) is questionable for EMP operating in low density plasmas. In accordance to Fig. (5) we conclude that the Eq. (5) is not satisfied because of the large values of the ratios $J_{rd}(T_w)/J_{eTh}(T_e)$. This is also the case for the electron temperature fitting of Fig. (6) which deviates from the values of T_e provided by other plasma diagnostics.

These results only involve the electrons emission and collection processes and are weakly dependent of the ion motion which is neglected in the previous derivation of Eq. (5). The classical model of Eq. (5) only considers that the thermionic and the plasma electron groups contributes to the probe current for $V_p \leq V_{sp}$. In order to explain the experimental findings discussed before, additional electron populations would be close to the surface of the EMP that also contribute to the probe current [5].

The physical origin of this third electron group would be the development of a plasma potential estructure around the EMP. In low density plasmas the large thermionic emission currents would modify the classical plasma sheath around the probe. These structures, also denominated *virtual cathodes*, might develop a plasma potential minimum that could trap and/or return a fraction of the emitted electrons by the EMP [12, 13, 14]. We conjecture that, despite the large ratios of $J_{rd}(T_w)/J_{eTh}(T_e)$, this electron confinement would limit the extent of the EMP perturbation in the nearby plasma and explains the reliable values of V_{sp} in low density plasmas.

In order to evidence the role of this third electron group the previous model for $V_p \leq V_{sp}$ based only on the emitted and plasma electron groups was refined in Ref. [5]. An additional electron group at the surface of the probe approximated by a Maxwellian distribution with an effective temperature. This parameter allows to fit the slopes of Fig. (6) and this result suggests that different electron groups are present at the surface of the electron emitting EMP. The consideration of a Maxwellian electron group is only regarded as an approximation and does not preclude other electron energy distribution functions for this electron group that might provide similar results.

This *floating point* method is more invasive than the *inflection point* discussed before which involves lower thermionic electron emission currents. However, in low density plasmas the noisy IV curves as those of Fig. (2) force the experimenters to operate the EMP in the high electron emission regime. The basis of the EMP operation in this latter case should be revised using more involved models that Eq. (5) to account for the different electron groups.

Acknowledgments

This work was funded by the Ministerio de Economía Ciencia e Innovación of Spain under Grant ESP2013-41078-R. The authors also acknowledge the partial support from Aernnova Engineering S.A. and from the Centro de Tecnologías Aeronáuticas, Vitoria (Spain). The work of S.P. Tierno was supported through the fellowship Grant FPU from the Ministry of Educación Cultura y Deporte of Spain.

References

[1] Goebel D M and Katz I (2008) *Fundamentals of Electric Propulsion: Ion and Hall Thrusters* (New York: John Wiley and Sons)
[2] Shun'ko E V (2009) *Langmuir probe in theory and practice* (Florida: Universal Publishers)
[3] Smith J R, Hershkowitz N and Coakley P (1979) Rev. of Sci. Instrum. **50** (2) p. 210
[4] Sheehan J P and Hershkowitz N (2011) Plasma Sources Sci. Technol. **20** (6) 063001
[5] Tierno S P *et al* (2013) IEEE Trans. Plasma Sci. **41** (4) p. 695
[6] Marek A *et al* (2006) Czech J. Phys. **36** (2) p. B932
[7] Schrittwieser R *et al* (2002) Plasma Phys. Control Fus. **44** (5) 567. Balan P *et al* (2003) Rev. Sci. Instrum. **74** (3) p. 1583
[8] Schrittwieser R *et al* (2009) J. Plasma Fusion Res. SERIES **8** p. 632
[9] Takamura S *et al* (2004) Contrib. Plasma Phys. **44** (1-3) p. 126. Ye M Y and Takamura S (2000) Phys. Plasmas **7** (8) p. 3457
[10] Conde L and León L (1994) Phys. Plasmas **1** (8) p. 2441
[11] Popov Tsv *et al.* (2009) Plasma Phys. Control. Fusion **51** 065014. Dimitrova M *et al.* (2014) Contrib. Plasma Phys. **54** (3) p. 255
[12] Gyergyek T *et al.* (2009) Plasma Sources Sci. Technol. **18** (3) 035001
[13] Sheehan J P *et al* (2013) Phys. Rev. Lett. **111** 075002
[14] Dias F M 1993 *Use of emissive probes in HF plasmas* (Microwave discharges: Fundamentals and Applications) eds. C.M. Ferreira and M. Moisan (New York: Plenum Press) p. 291
[15] Kusaba K and Shindo H (2007) Rev. of Sci. Instrum. **78** (12) 123503
[16] Raizer Y P (1991) *Gas Discharge Physics* (Berlin: Springer-Verlag) pp. 103-107
[17] Langmuir I (1916) Phys. Rev. **7** (3) p. 302
[18] Criado E *et al* (2012) Phys. Plasmas **19** (2) 023505
[19] Tierno S P *et al* (2013) Contrib. Plasma Phys. **53** (1) p. 16.

Sticking non-stick: Surface and Structure control of Diamond-like Carbon in Plasma Enhanced Chemical Vapour Deposition

B J Jones[1,2] **and N Nelson**[3]

[1] School of Materials, University of Manchester, Oxford Road, Manchester M13 9PL UK
[2] School of Applied Sciences, University of Huddersfield, Queensgate, Huddersfield, West Yorkshire HD1 3DH UK
[3] School of Engineering and Design, Brunel University, Kingston Lane, Uxbridge, Middlesex UB8 3PH UK

E-mail: b.j.jones@physics.org

Abstract. This short review article explores the practical use of diamond-like carbon (DLC) produced by plasma enhanced chemical vapour deposition (PECVD). Using as an example issues relating to the DLC coating of a hand-held surgical device, we draw on previous works using atomic force microscopy, X-ray photoelectron spectroscopy, Raman spectroscopy, scanning electron microscopy, tensiometry and electron paramagnetic resonance. Utilising data from these techniques, we examine the surface structure, substrate-film interface and thin film microstructure, such as sp^2/sp^3 ratio (graphitic/diamond-like bonding ratio) and sp^2 clustering. We explore the variations in parameters describing these characteristics, and relate these to the final device properties such as friction, wear resistance, and diffusion barrier integrity. The material and device characteristics are linked to the initial plasma and substrate conditions.

1. Introduction

A low-friction, hard-wearing, biocompatible, barrier thin-film material, diamond-like carbon (DLC) contains both sp^2 (graphitic) and sp^3 (diamond-like) bonded carbon atoms, and often a significant level of hydrogen. Medium-range order, clustering of the sp^2 bonded atoms within a sp^3 matrix, plays a substantial part in the properties of the material. Tuning the balance of the constituents, phases and ordering allows the creation of a material with properties that can be tailored to match requirements resulting in application in many areas including electronics [1], aerospace [2], and medical devices [3] as well as beer storage [4].

Plasma enhanced chemical vapour deposition (PECVD) is a widely used laboratory technique for DLC deposition. This work reviews research utilising micro and nanoscale characterisation techniques in order to understand the effects of production processes and improve film quality and function for the end user. Application of methods such as electron microscopy (SEM), atomic force microscopy (AFM), X-ray photoelectron spectroscopy (XPS) and X-ray diffraction (XRD) has enabled improved understanding of the changes in film structure, surface and properties with variation of plasma conditions. This includes work to improve film adhesion [5], create machining tools with improvement in lifetime and reduction in power consumption [6], and enhance the thin film diffusion and surface properties to facilitate production of improved medical tools and implants [3,7].

This short review paper examines factors affecting PECVD deposited DLC structure and properties by exploring some issues associated with the coating of surgical device, figure 1 [3,7].

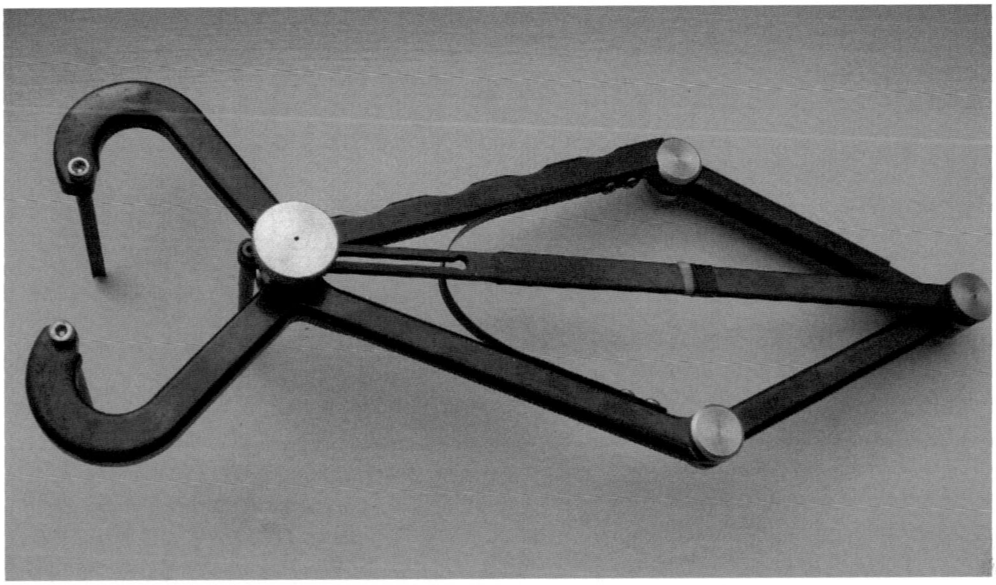

Figure 1. Diamond-like carbon coated aluminium surgical assist device. Reprinted from *Diamond and Related Materials* [7] copyright (2010) with permission from Elsevier.

2. Plasma control

The rf power supplied, or negative self bias induced, may be the primary method within a single PECVD system with which to control the deposition energy of ion species. By increasing the bias voltage the film structure can be adjusted from hydrogen-rich polymeric coating, to sp^3-rich diamond-like carbon to a more graphitic-like carbon. The optimum structure will be related to the application, the different films have varying defect densities and diffusion barrier properties, as well as friction and wear characteristics [1,2,7,8]. For example, Zolgharni et al. [6] use an Ar/C_2H_2 precursor and show an optimum bias voltage of 400-550V for friction and wear in pin-on-disc tribological tests, and 450V for coating of drill bits used for dry machining of aluminium. For the surgical device, a lower bias of 200V is optimal for diffusion barrier properties [7].

However, the ion energy and species distribution are also affected by factors such as precursor gas mix, chamber pressure, and substrate or cathode structure [9-11]. For example, varying the Ar/C_2H_2 balance of the precursor gases affects the film surface structure, as shown by AFM analysis where average roughness increases from 2.6nm to 5.0nm and peak density decreases from approximately 50 to approximately 12 peaks per square micron, as the argon content of the precursor gas mix is raised from 0 to 45% [11] as shown in figure 2. In addition the clustering of the sp^2 regions increases with increasing Ar in the precursor mix, as observed by electron paramagnetic resonance (EPR), leading to an increase in film conductivity [11]. It is postulated that the changes in the film structure are related to the balance of impinging species affecting film nodulization and sp^2/sp^3 ratio related to the balance of surface diffusion and subplantation and relaxation of internal stresses [11-13]. These structural changes lead to an optimal argon content for a particular application, relating to properties such as friction or material adhesion in dry drilling [6,11] or adherence of biofilms for medical devices [3,7].

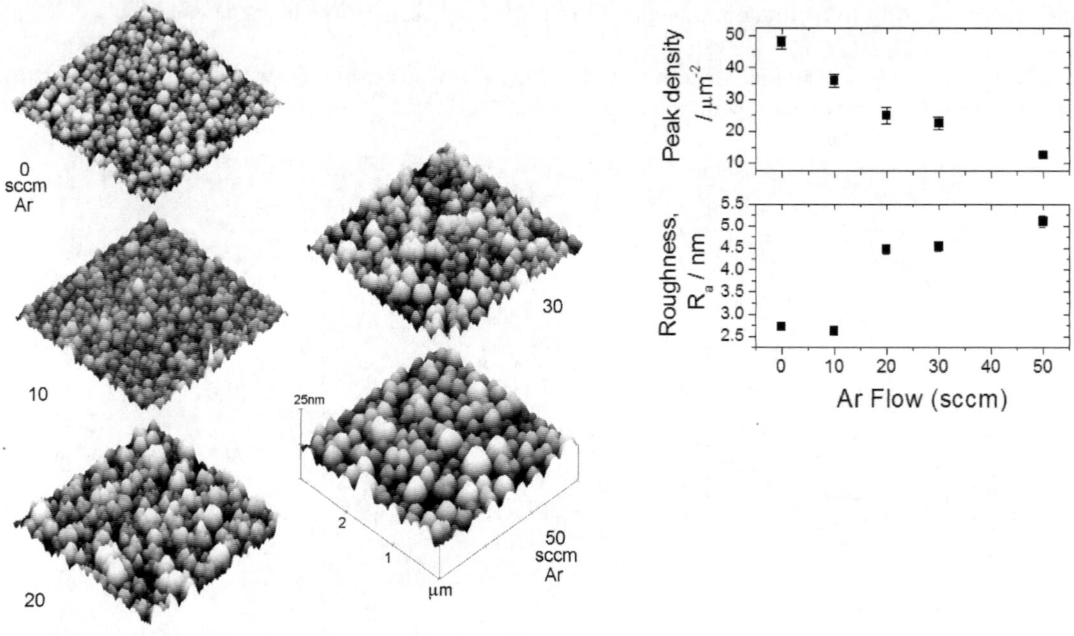

Figure 2. Atomic force microscopy images and calculated texture parameters of DLC on silicon showing changes to the surface structure with increasing argon content of precursor gas mix. Reprinted from *Diamond and Related Materials* [11] copyright (2008) with permission from Elsevier

The film chemistry and structure can be affected by incorporation of species via the precursor gas mix [3,7,14,15], or post treatment processes [1,16,17]. For the medical assist instrument [7], in addition to limiting adherence of biofilms, the primary purpose of the DLC was for diffusion barrier properties, to protect the instrument from biofluid corrosion and to protect the patient from exposure to aluminium. Barrier properties were optimised by incorporating silicon into the film, through the inclusion of tetramethylsilane (TMS) in the precursor gas mix. Aggressive erosion tests through immersion of test pieces in sodium hydroxide for 30mins showed a precursor ratio TMS: C_2H_2 of 1:4 optimal for barrier properties, resulting in a 53% reduction in aluminium transfer against an uncoated sample, and 45% reduction versus an unsilicated amorphous carbon. The results are also correlated with an increase in water contact angle, reflecting the behaviour of fluid at the film surface [7] and are summarised in figure 3. The incorporation of silicon to improve the diffusion barrier properties is consistent with that seen by other groups [14,15] and may be related to increased hydrogen and sp^3 fractions in the film [14,5].

3. Substrate effects

Factors relating to the substrate also affect the flux, energy, species and interactions of depositing ions and strongly affect the resultant film structure. Although PECVD is a non-line of sight process, and has been described as producing uniform films over large areas, the structure of many devices to be coated, such as the medical assist device [7] are not uniformly planar and may perturb the plasma, leading to variations in the deposited film. Nelson et al [9] show how substrate geometry may affect the DLC film, by exploring variation in the deposited roughness, R_q of 307±47 nm over 10x10 µm, compared to 29.4±2.7 nm for the film deposited on steel, film with the orientation of the substrate, using silicon as a model surface.

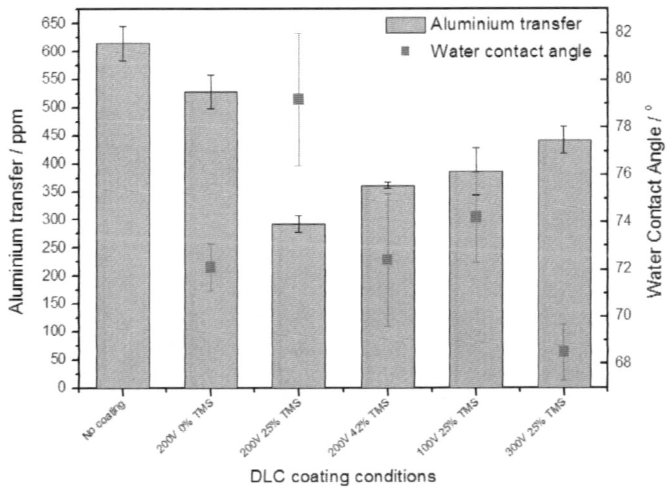

Figure 3. Effect of silicon content of precursor (TMS) on film barrier properties, characterised by transfer of aluminium in NaOH immersion, and water contact angle. Data from [7].

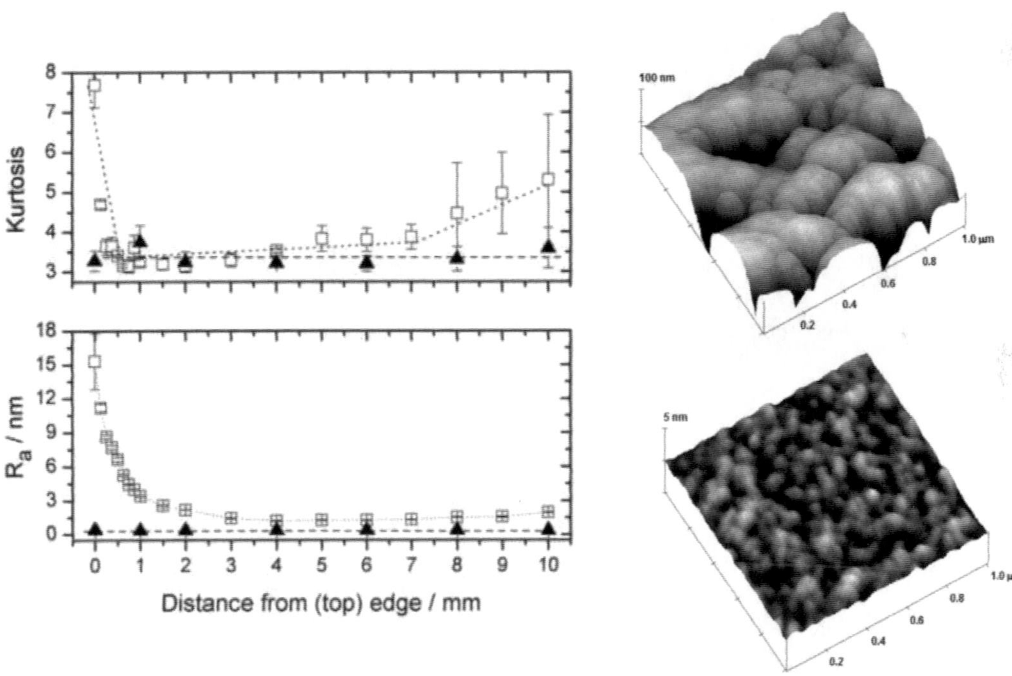

Figure 4. Effect of substrate orientation and geometry on DLC surface texture. Surface parameters and example AFM images for vertically orientated surface (top right, □) and horizontally orientated surface (lower right, ▲). Note the difference in z-scale. Reprinted from *Surface and Coatings Technology* [9] under creative commons http://creativecommons.org/licenses/by/3.0/legalcode

They show variation in film roughness (R_a) of a factor of over 30, increasing from 0.4nm on a horizontal sample, to 15nm at the top of a vertically positioned sample, figure 4. Film thickness varies by a factor of 2 and variation is exhibited in the nodularity and sp^2/sp^3 ratio, assessed through Raman spectroscopy, over a lateral or vertical scale of approximately 10mm. This is linked to the inhomogeneous distribution of ion species within the plasma and to field concentration at the sharp vertical edge, affecting plasma sheath formation and shielding effects, leading to variation in effective energy, flux and species of the impinging ions, and therefore to variation in surface diffusion and subplantation mechanisms in the film growth [9]. These factors ultimately affect the friction and critical load of the device [9].

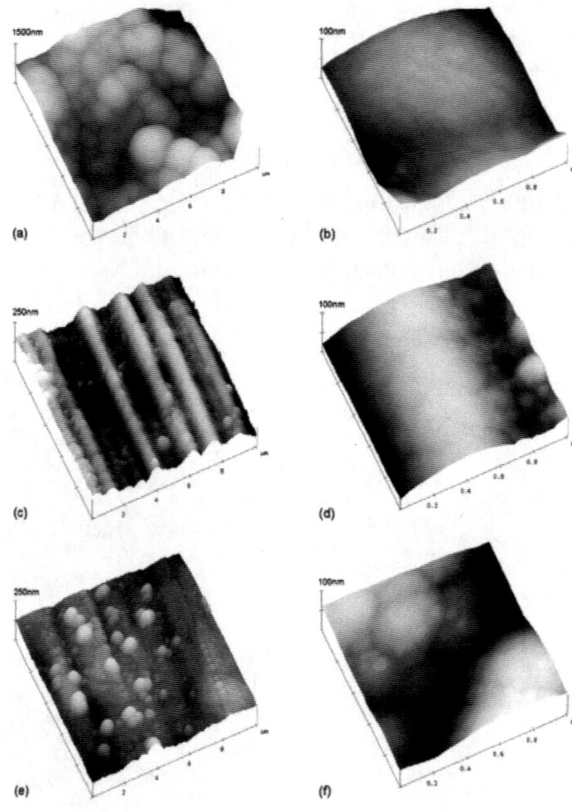

Figure 5. DLC produced with the same nominal parameters on (a,b) epoxy resin and (c,d) concurrently produced on stainless steel and (e,f) separately produced on stainless steel, over (a,c,e) 10x10μm and (b,d,f) 1x1μm, showing variation of DLC structure with substrate material. Copyright © 2012 John Wiley & Sons, Ltd. Reproduced from [18], with permission.

Atomic force microscopy studies of DLC deposited on epoxy resin in comparison to steel for application to aircraft landing gear [2,18] clearly show the differences between nominally similar coatings on radically different substrates, figure 5. Although the DLC deposition conditions are similar to those optimized for machining of aluminium [6] the film on epoxy has a root mean square

Journal of Physics: Conference Series **768** (2016) 012011 doi:10.1088/1742-6596/768/1/012011

the skewness and kurtosis of these surfaces also exhibit differences, but perhaps the most significant change in surface texture is indicated by the peak density, 0.12 μm^{-2} on epoxy compared to 1.1 μm^{-2} on steel, reflecting differences in the surface diffusion of the depositing species [18].

The chemistry of the film surface is also altered. The sp^2/sp^3 ratio can be calculated via Raman spectroscopy, as above, [9], however, other works utilise XPS to examine the surface microstructure. Analysis techniques may focus on C1s deconvolution [5,18] or X-ray excited auger spectroscopy measurements [18,19]. XPS results show that the sp^2 fraction of the film on epoxy is higher than that of the DLC on steel, by 3-8 abs.%, though the exact figure depends on calculation procedure [18]. These differences in the film structure, in addition to the variation in modulus and interface adhesion, lead to variation in the device performance, the DLC+Epoxy combination performing better in wear and friction tests for applications to aircraft landing gear [2,18].

The presence of substrates with a high degree of volatility, or which are easily sputtered, also affects the gas mix of the plasma, and subsequent film structure. The research on DLC coated on epoxy and steel [18] demonstrates transfer of Ca, Na, N and O from the epoxy resin structure into the DLC film deposited on steel within the same deposition period, suggesting transfer of components of the epoxy to the gas phase through heating, sputtering or outgassing, and subsequent deposition with the DLC [18].

4. Interface region

The adhesion of the DLC film to the substrate material or device is of high importance, failure of the film adhesion can lead to catastrophic failure of the device, for example in a biomedical implant. Although the film is biocompatible [20], film failure may lead to an adverse reaction in the patient to the underlying metal surface, or enhancement of aseptic loosening process of an implant as a result of debris from the film [3], or for some titanium-DLC films, debris may be internalised in bone marrow [21]. The epoxy structure discussed above is an example of a thick interlayer that improves adhesion [2], many groups have utilised interlayers to improve adhesion of diamond-like carbon to a substrate, this may be metallic such as tungsten [22] or, often in adherence to steels, based on a graduated silicon layer [5,6,9,11]. In addition the substrate cleaning process is highly important, and many teams use an argon plasma pretreatment step; however, earlier work [5] highlights that the pretreatment conditions vary significantly in terms of power and duration. The argon plasma treatment step has a dual purpose, at low power the substrate is cleaned through the removal of hydrocarbon contamination [5]; however, in a study on 304 stainless steel substrates [5] XRD and XPS analysis, coupled with electron microscopy, show that at higher powers the argon plasma processing restructures the metal surface, preferentially removing chromium from the top layer and generating a nanotextured, nanocrystalline surface, figure 6.

This processing improves film adhesion in bend tests by approximately three orders of magnitude [5]. However, the DLC structure is also affected by the changes to the substrate, as suggested in section 3 above. In this instance the rms surface roughness, R_q is reduced, from 37 nm to 21 nm and the nodularity of the surface is altered, as shown in figure 7. The sp^2 fraction of the DLC surface is also affected by the microstructured interface [5]. Failure to adequately pretreat the substrate surface can lead to rapid film delamination, figure 8.

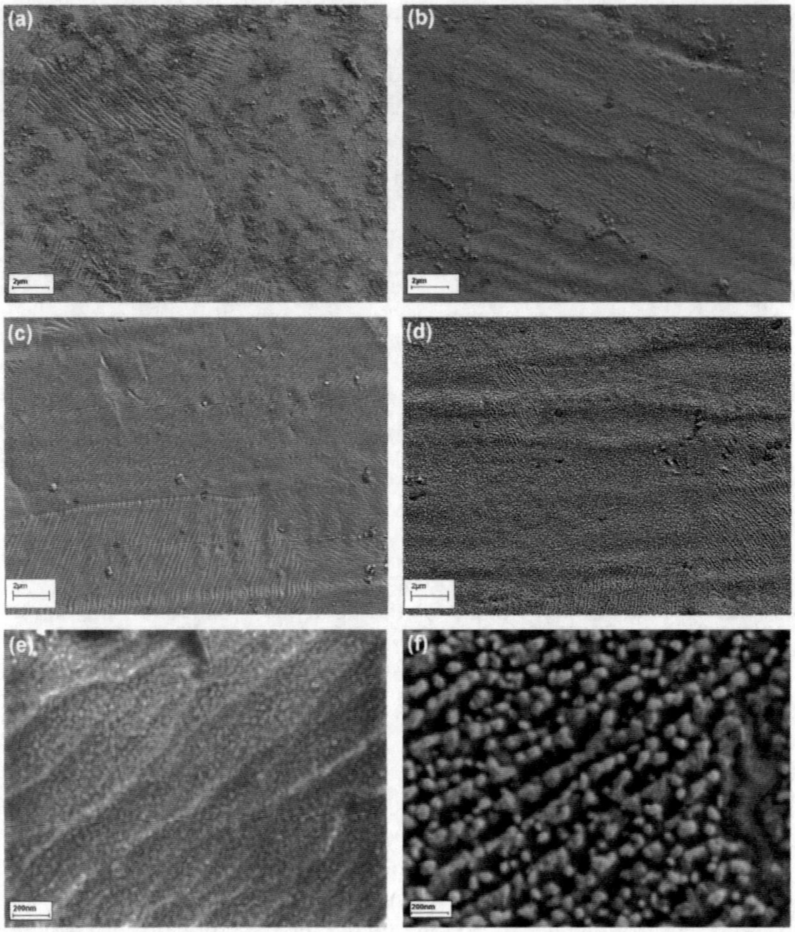

Figure 6. 304 Stainless steel with a) sonication preclean and argon pretreatment with bias voltages of b) 300V c) 450V and d) 600V. Higher magnification images of e) 450V and d) 600V reveal nanotextured surface. Reprinted from *Diamond and Related Materials* [5] copyright (2011) with permission from Elsevier

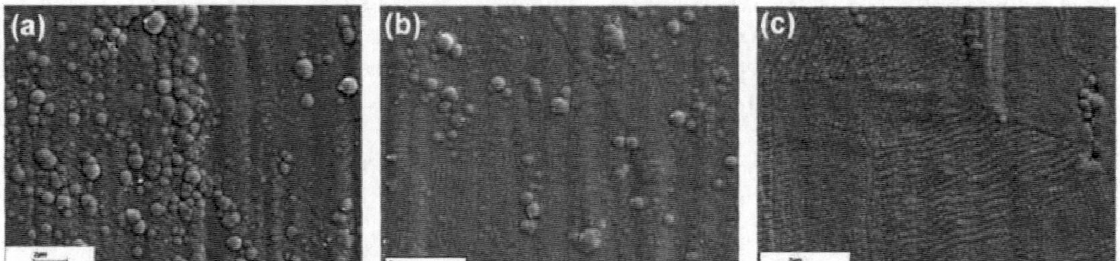

Figure 7. Scanning electron microscopy images of DLC films deposited on substrates with argon plasma pretreatment of a)300V, b) 450V and c)600V showing effects of substrate structure on DLC texture. Reprinted from *Diamond and Related Materials* [5] copyright (2011) with permission from Elsevier.

5. Conclusions

In this short review of previous work, issues affecting diamond-like carbon produced by plasma enhanced chemical vapour deposition have been briefly examined. Deliberate control of the plasma conditions, for example through rf power or negative self bias voltage, precursor gas mix or pressure, may cause variation in depositing in species or energy, leading to changes in the surface interaction mechanism and variation in the film structure and chemistry. In addition, other factors, such as substrate material, outgassing, contamination, and substrate shape, can lead to significant variation in effective energy and flux of impinging ions, as well as predominant ion species. These factors strongly influence film production, film structure and resultant properties, such as integrity of diffusion barrier, friction, wear and adhesion. Substrate interactions are therefore important to consider in DLC production and also have relevance to other PECVD materials such as silicon dioxide based films.

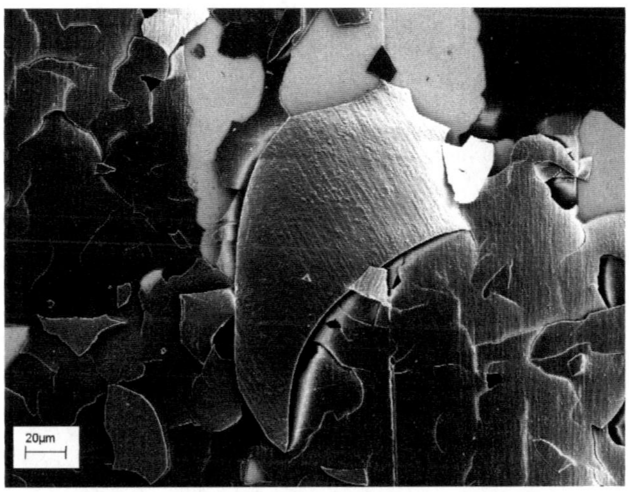

Figure 8. Unstuck non-stick: Failure to adequately prepare the substrate leads to film delamination.

Acknowledgements

Thanks to Lorna Anguilano, Tony Anson, Bob Barklie, Bob Bulpett, Joe Franks, Arun Mahendrun, Jesus Ojeda, Sander Podgoric, Richard Rakowski, Alan Reynolds, and Massoud Zolgharni. We gratefully acknowledge support of the Engineering and Physical Sciences Research Council, Technology Strategy Board, The Royal Society and the University of Huddersfield.

References:

[1] B.J. Jones, R.C. Barklie, G. Smith, H. El Mkami, J.D. Carey and S.R.P. Silva *Diamond and Related Materials* **12** (2003) 116
[2] S. Podgoric, B.J. Jones, R. Bulpett, G Troisi, J. Franks *Wear* **267** (2009) 996
[3] B.J. Jones and A.W. Anson *Materials World* **16**:8 (2008) 24
[4] C. Casiraghi, J. Robertson, A.C. Ferrari *Mater. Today* **10** (2007) 44
[5] B.J. Jones, L. Anguilano, J.J. Ojeda *Diamond and Related Materials* **20** (2011) 1030
[6] M. Zolgharni, B.J. Jones, R. Bulpett, A.W. Anson, J. Franks *Diamond and Related Materials* **17** (2008) 1733

[7] B.J. Jones, A. Mahendran, A.W. Anson, A.J. Reynolds, R. Bulpett and J. Franks *Diamond and Related Materials* **19** (2010) 685
[8] Robertson J. *Materials Science & Engineering R- Reports* **37** (2002) 129
[9] N.Nelson, R.T. Rakowski, J.Franks, P. Woolliams, P. Weaver, B.J. Jones *Surface and Coatings Technology* **254** (2014) 73
[10] Peng XL, Barber ZH, Clyne TW. Surface roughness of diamond-like carbon films prepared using various techniques. *Surface and Coatings Technology* **138** (2001) 23.
[11] B.J. Jones, S. Wright, R.C. Barklie, J. Tyas, J. Franks, A.J. Reynolds *Diamond and Related Materials* **17** (2008) 1629
[12] Y. Lifshitz, G.D. Lempert, E. Grossman *Phys. Rev. Lett.* **72** (1994) 2753
[13] C. Riccardi, R. Barni, M. Fontanesi, P. Tosi *Chem. Phys. Lett.* **329** (2000) 66
[14] P.D.Maguire, J.A.McLaughlin, T.I.T.Okpalugo, P.Lemoine, P.Papakonstantinou, E.T. McAdams, et al., *Diamond Relat. Mater.* **14** (2005) 1277
[15] C.-P. Klages, A. Dietz, T. Höing, R. Thyen, A. Weber, P. Willich, *Surf. Coat. Technol.* **80** (1996) 121
[16] R.U.A. Khan, J.D. Carey, S.R.P. Silva, B.J. Jones and R.C. Barklie *Physics Review B* **63** (2001) 121201(R)
[17] B.J. Jones, R.C. Barklie, R.U.A. Khan, J.D. Carey and S.R.P. Silva *Diamond and Related Materials* **10** (2001) 993
[18] B.J. Jones, J.J. Ojeda *Surface and Interface Analysis* **44** (2012) 1187
[19] A. Mezzi, S. Kaciulis, *Surf. Interface Anal.* **42** (2010) 1082
[20] L.A. Thomson, F.C. Law, N. Rushton, J. Franks *Biomaterials* **12** (1991) 37
[21] A. Bruinink, A. Schroeder, G. Francz, R. Hauret *Biomaterials* **26** (2005) 3487
[22] T. Takeno, T. Sugawara, H. Miki, T. Takagi, *Diamond Relat. Mater.* **18** (2009) 1023

Visualisation of plasma induced processes in interelectrode gap of micropinch installation using laser illuminator on molecular nitrogen

E Filippov, S Sarantsev, I Raevsky, A Savjolov
National research nuclear university "MEPhI", Kashirskoe shosse 31, Moscow, Russia

E-mail: jiienechka@gmail.com

Abstract. The paper presents the results of researched processes in the interelectrode gap of High-current low-inductance vacuum spark (HLIV) on the "PION" system [1] by means of active laser diagnostics with visualisation of radiation field. The influence of the cathode geometry and position of the trigger system on the plasma dynamics of HLIV is considered. The regularities of the discharge for different initial conditions are established. The conditions under which the most stable plasma point is formed in space and time are determined.

1. Introduction

Micropinch discharge (*MPD*) is usually known as pulsed discharge with a high aspect ratio, which forms a short-lived plasma in very small volume with high temperature and density ($Ne \geq 10^{20}$ cm^{-3}, $T > 1$ keV). One can underline low-inductive vacuum spark (LIVS) among the setups that implement micropinch mode where micropinch is usually known as plasma point. Installations which are based on low inductive vacuum spark (LIVS) are used for a set of applied research and technical tasks as sources of electromagnetic radiation microwave, infrared, visible, ultraviolet and x-ray ranges; as well as powerful plasma flows [2]. Micropinch discharge is a source of vacuum ultraviolet and soft x-ray radiation and is promising for the purposes of X-ray [3], the creation of lasers, X-ray range [4] etc; development of effective sources of multiply charged ions for nuclear physics experiment in terms of technical applications [5]. As the features of installations based on LIVS one can notice the simplicity of the design, the ability to work in frequency mode, the long life time, low cost, and that the physical processes occurring in various Z - pinch discharges, have a lot in common.

In the study of the dynamics of micropinch discharge from the moment of its initiation to the stage of pinching and collapse inside entire section of the discharge gap, it requires application of a set of diagnostic tools, allowing to diagnose the plasma in a wide range of parameters. Methods of diagnostics are easier and more logical to try in small installations, for example, LIVS at relatively low values of the stored energy (0.5 - 10 kJ). At the same time an important thing is the absence of contact with the plasma, so optical diagnostic techniques as passive - based on registration of intrinsic luminosity of plasma and active - on the basis of laser sensing have significant sense.

This paper is the research of micropinch using laser on molecular nitrogen.

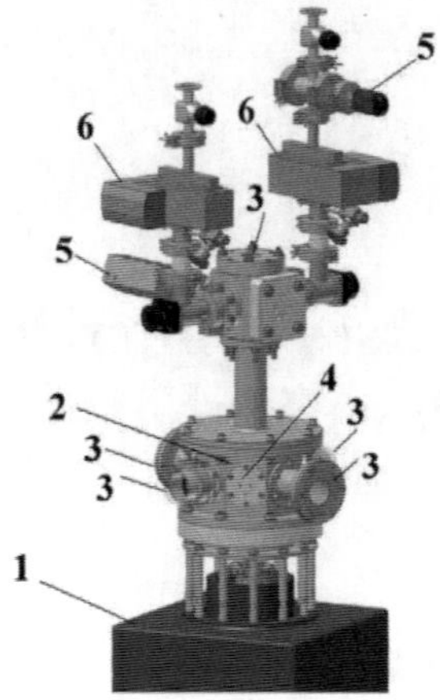

Figure 1. Installation appearance: 1 – capacitor (PC); 2 – vacuum chamber; 3 – diagnostic window; 4 – vacuum inputs for auxiliary triggers; 5 – pressure sensors; 6 – turbo molecular pump

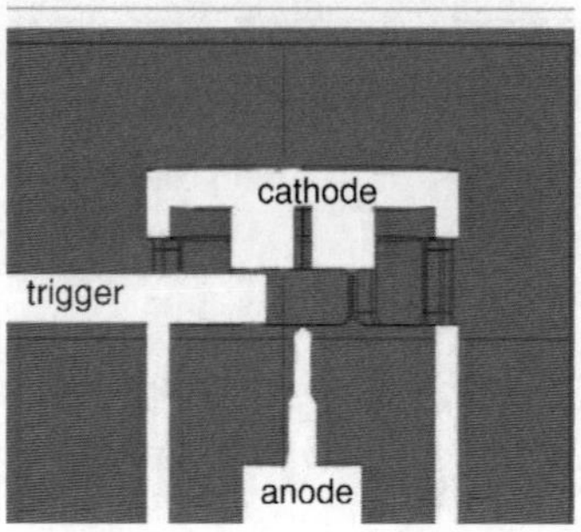

Figure 2. Sample of electrode configuration

2. Experimental setup and results

The work was performed at the "PION" installation with such type of a plasma source as high-current low-inductive vacuum spark (HLIVS). Main parameters are the follows: stored energy is up to 2 kJ and period of discharge is 5.6 ms. Appearance of the installation is presented in figure 1. Electrodes of «PION» installation are the cathode of various shape and peak anode with diameter D = 3 mm (figure2). Material of the electrodes is Fe (Steel №45, purity 97%, polished). Distance between electrodes during the experiment varied from 3 to 9 mm. Voltage on the main capacitor battery (capacity C = 12 μF) was up to 20 kV, the pressure in the chamber was equal to 10^{-6} Torr. Probing was performed by a laser pulse through diagnostic quartz windows, assigned from the discharge axis at a distance of 350 mm to prevent ingress of dust products after discharge. Initiation of the discharge was committed by a source of erosive plasma type (the breakdown on surface of ceramic insulator). One of the Thepler methods [6] was used to investigate the influence of shape of the electrodes on the plasma dynamics of LIVS, and LGI-21 laser on molecular nitrogen was applied as illuminator.

The experimental scheme is shown in Figure 3a. Light beam expander was assembled on base of lenses L_1 and L_2 to increase the laser beam diameter from 3 to 12 mm. Mirror M_1 reflects part of the laser radiation and redirects it to the PEC (coaxial photocell) to determine the exact moment of laser operation. Optical system consisting of two quartz lens L_3 (f = 1 m) and L_4 (f = 0.25 m) and three interference mirrors with a reflection coefficient 99.99% on the laser wavelength and located at 45° M_2, M_3 and M_4 were used to make an optically conjugated image of the electrode gap on a matrix camera with magnification M = 2. Application of this scheme allowed removing the camera at a distance of about 6 m from the discharge, which significantly reduced the ambient light from the plasma and reduced the influence of electromagnetic interference on the recording apparatus. In front of the camera pinhole with orifice of 300 microns and an interference filter on the wavelength λ = 337 nm with a bandwidth of 6 nm have been located. Image Registration was implemented on a camera Canon EOS 1000D with removed UV filters and objective lenses in front of the matrix. Micropinch formation was recorded by the current waveform obtained from the Rogowski coil, and a flash of X-rays, registered using the pin-diode (figure 8). So the synchronization between the discharge pulse and laser pulse was established via delay of the pulse generator GZI-21 between the input signal of Rogowski coil and the output signal for laser probe beam (figure3b).

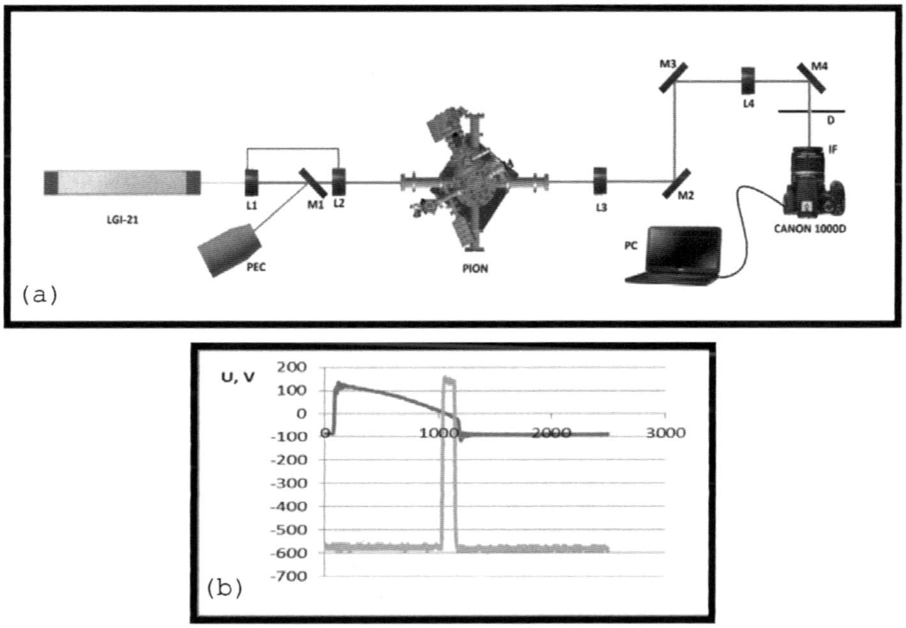

Figure 3. (a) Scheme of experiment on «PION» installation, (b) Synchronization between the discharge pulse and the laser pulse. Red – GZI-21 and green – pulsed generator G5-15 that launches laser probing. Voltage is given in p.u.

Research was conducted for the three electrode systems, differing from each other by shape of the cathode (figure 4): (a) a cylinder with diameter 22 mm; (b) a cylinder of diameter 22 mm with orifice of 3 mm along an axis; (c) a cone with a base of 10 mm diameter and an opening angle of 90°. Bar of 3 mm diameter with spike was used as anode in all cases. Trigger was located on the side of the electrodes (Model presumes that total voltage of the battery is applied instantly in the electrodes and in

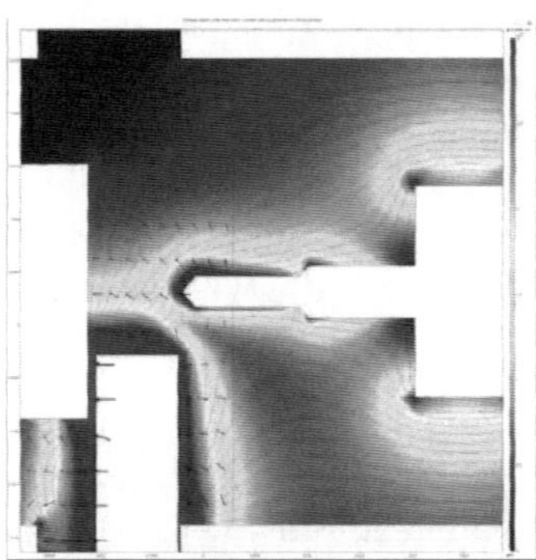
(a)

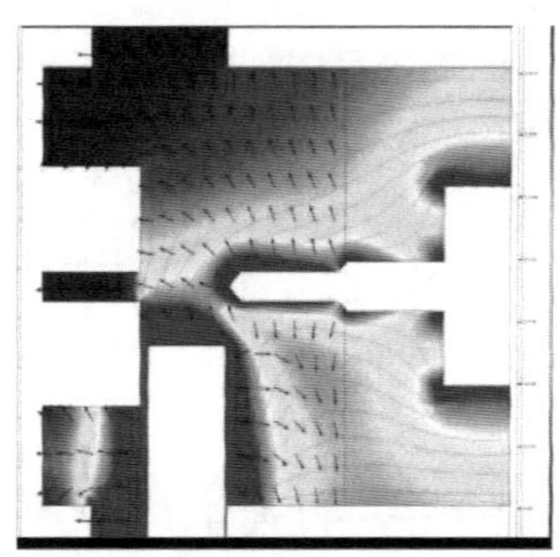
(b)

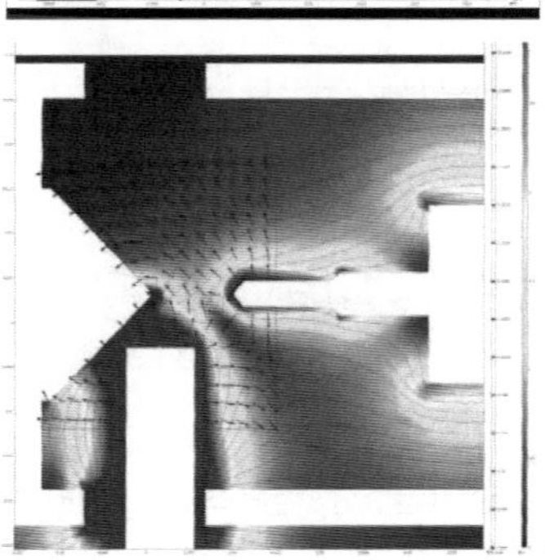
(c)

Figure 4. Variants of electrode systems used on the "Pion" setup. Colour – electric field lines. Red – maximum. Lines are equipotential lines of field.

the same time when trigger pulse occurs; simulating has been done in Comsol Multiphysics). For each configuration evolution of micropinch has been obtained using experimental shadows. Each separate shadow corresponds to one separate discharge, so it required big statistics.

After submitting of high-voltage pulse to the trigger electrode system, breakdown occurs on the dielectric surface (we observed it just on oscillograms by the delay between signals of discharge and trigger requiring for production of plasma) and auxiliary plasma extends to the electrode gap and initiates vacuum breakdown of the basic discharge. Formed as a result of the breakdown electrons are accelerated by an electric field towards the anode. So plasma cloud near the anode surface is formed, and it starts moving towards the cathode (figure5). In the first case we see it happens in more late moment – 400 ns where 0 ns is time when X-ray pulse happens (So time was negative if the laser pulse occurred before X-ray, conversely – positive). Probably it was because concentration of auxiliary plasma was high and it was difficult to turn it apart.

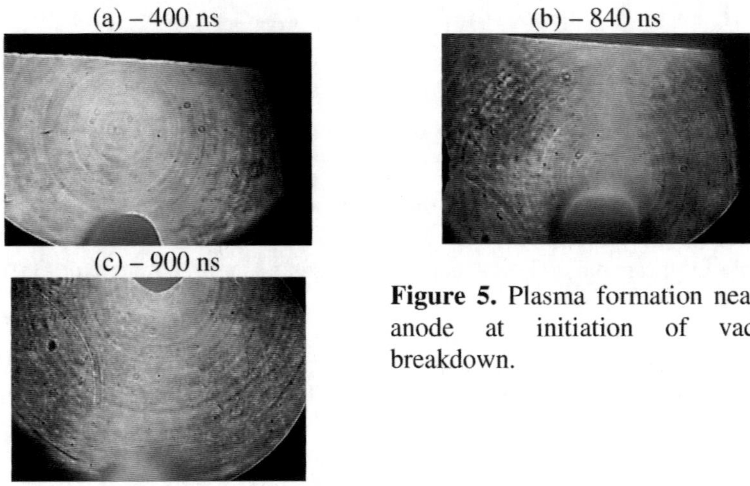

Figure 5. Plasma formation near the anode at initiation of vacuum breakdown.

Moving toward the cathode, anode plasma cloud begins to shrink to the discharge axis by magnetic forces of the discharge current, leading to the formation of the sausage-type instability. Depending on the shape of the electrodes necking may be offset from the axis of the discharge gap, or placed on its axis. In the figure 4 modeling of electrical potential and lines of electric field at initial conditions are presented. It was noticed that in the first and second cases we have more significant changes of field and so we could expect that there is the dependence discharge formation on localization of trigger here. Obtained shadows confirm that assumption for this case of flat cathode (figure 6).

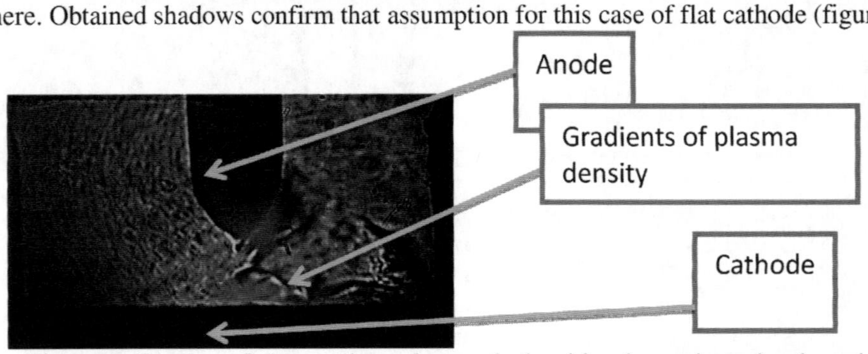

Figure 6. Curving of plasma channel towards the side where trigger has been located, t = - 0.1 μs

After necking, according to the theory of radiative collapse [7] in terms of Pease-Braginskii condition (discharge current must exceed a critical value of the current - 50 kA for iron), the plasma point (PP) is formed. When using a cylindrical cathode without orifice (figure 7a) the following

Journal of Physics: Conference Series **768** (2016) 012012 doi:10.1088/1742-6596/768/1/012012

features were observed: in a new system of electrodes (up to 120 shots) only one PP formed at time ~ 800 ns (60%) with scatter of 150 ns or - ~ 1100 (~ 25 %) with 100 ns; while amount of discharges increases, x-ray radiation moment shifts to 850 ns from the beginning of the current flow, and the dispersion increased to 300 ns. Localization region in the space is constricted of ~ $0.4 \times 1\ \text{mm}^2$ and displaced towards the trigger at distance 1 mm approximately. Two plasma points (figure 8) were formed unlikely (less than~ 15%), probably because of bridging currents of peripheral plasma. Herewith, time between peaks is between 10 and 50 ns.

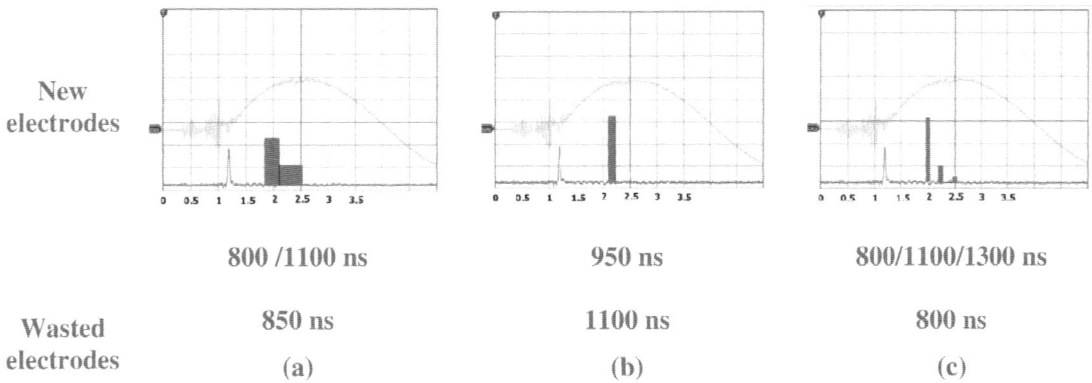

Figure 7. Dispersion of X-ray emission time and its probability (that matches amplitude in the figure) for three electrode systems (red rectangular bar): (a) - cathode without orifice, (b) - cathode with orifice, (c) - conical cathode. Red single pulse on all figures corresponds to X-ray emission in the beginning of the discharge evolution. Green – discharge current.

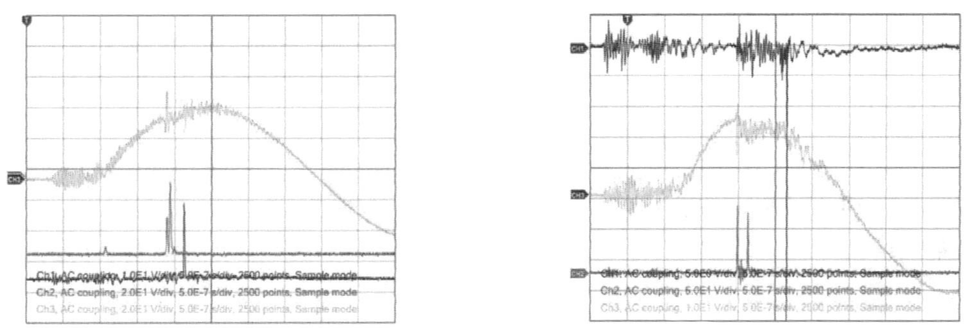

Figure 8. Waveform with double X-ray pulse for the cases with flat cathode. Red – X-ray, Blue – laser, Green – Discharge current. On the left - cathode without an orifice, on the right - cathode with an orifice

In the case of a cathode with an orifice (figure 7b) moment of X-ray emission also was changed while number of shots increases, but the origin of the change was different: after installing new electrodes (up to 120 shots) one PP mainly formed at time ~ 950 ± 80 ns (~ 80 %); then it was more likely to occur in later times, and after about 200 discharges micropinch formed mainly in time 1100 ± 60 ns from the beginning of the current flow. Localization region in the space is constricted by ~ $0.6 \times 1\ \text{mm}^2$, and displaced towards the trigger. Dual X-ray peaks were detected more often and time interval between two stages of maximum compression was more significant than in the first case when cathode without orifice is used (150 - 200 ns).

Using the conical cathode (figure 7c) plasma point formed in moment of time 800 ns ~ (~ 80%) or about 1100 ns (~ 15%) or about 1300 ns (~ 5%) with a dispersion of no more than ~ 80 ns. When wasted electrodes were investigated, probability of later moments was reduced to zero. After 200 shots micropinch preferably was formed at the time of 800 ns from the start of current flow. In this case, the spatial domain of preferential formation of instability (size ~ 0.4 × 0.4 mm^2) located symmetrically relative to the axis of the discharge gap.

3. Conclusion

The experimental results appear to try to explain physical processes, occurring in the interelectrode gap of micropinch installation. Using the method of a laser probing information about the dynamics of LIVS plasma for three different electrode configurations has been obtained experimentally. Such system of registration with digital camera is very convenient and one can obtain good quality of shadows spending short time. The most significant influence on the localization of the hot point was observed in the case of a flat cathode, so the region is shifted relative to the system axis. Hereby, in the case of conical electrodes the spatial domain is located symmetrically relative to the axis of the discharge gap in smaller volume and formation of PP is more stable.

While amount of discharges increases moment of x-ray radiation shifts in all cases but behavior of this displacement has different character. The dependence the time interval between the formations of two PP on the configuration of the electrode system also can verify different origin so the electrode system has significant role in micropinch processes.

References

[1] Bashutin O, Vovchenko E, Dodulad E, Raevskiy I, Saveliev A, Stepanenko A M 2011 Effect of initiation conditions on low inductance vacuum spark plasma dynamics and processes at the electrode surface *XIV All-Russian Conference "Diagnostics of high-temperature plasma" DVP-14*

[2] Dolgov A, Salakhutdinov G 2003 Material transport in a Low-Inductance vacuum *Plasma Physics reports* **29** 757-763

[3] Dolgov A, Zemenchekova N, Prokhorovich D 2011 Experimental evidence of electron runaway *Plasma Physics Reports* **T 37** 227–232.

[4] Jen Hun Li 1999 Dynamics of micropinch discharge of low-inductive vacuum spark type with transverse plasma ignition Phd dissertation (Moscow: MEPhI) 125

[5] Veretennikov V, Gurey A, Dolgov A, Semenov O, Tihomirov A 1995 Micropinch plasma discharge as a source of heavy elements *JETP Letters* **21** 78-81

[6] Vasiliev L 1968 *Shadow methods* (Moscow: Nauka) 400

[7] Koshelev K N, Pereira N R 1991 Plasma points and radiative collapse in vacuum sparks *J. Appl. Phys.* **69**

Barrier SiO_2-like coatings for archaeological artefacts preservation

M Prochazka, L Blahova and F Krcma

Faculty of Chemistry, Brno University of Technology, Purkynova 464/118, Brno 612 00, Czech Republic

xcprochazkam@fch.vutbr.cz

Abstract. Thin film chemical vapour deposition technique has been used for more than 50 years. Introducing organo-silicones as precursors, e.g. hexamethyldisiloxane (HMDSO) or tetraethyl orthosilicate (TEOS), brought new possibilities to this method. Barrier properties of thin films have become an important issue, especially for army and emergency services as well as for food and drink manufacturers. Our work is focused on protective HMDSO thin films for encapsulating cleaned archaeological artefacts, preventing the corrosion from destroying these historical items.Thin films are deposited via plasma enhanced chemical vapour deposition (PECVD) technique using low pressure capacitively coupled pasma in flow regime. Oxygen transmission rate (OTR) measurement was chosen as the most important one for characterization of barrier properties of deposited thin films. Lowest OTR reached for 50 nm thin film thickness was 120 cm3 m-2 atm-1 day-1. Samples were also analyzed by Fourier Transform Infrared spectrometry (FTIR) to determine their composition. Optical emission spectra and thin film thickness were measured during the deposition process. We optimized the deposition parameters for barrier layers by implementation of pulsed mode of plasma and argon plasma pre-treatment into the process.

1. Introduction

Chemical vapour deposition has been used for thin film production for more than 50 years. We can find many successful applications in different industry branches, e.g. in automotive, food industry or in medicine. During 1980's, this deposition technique has been improved by adding plasma to the process. Plasma enhanced chemical vapour deposition (PECVD) or plasma assisted chemical vapour deposition (PACVD) allow using many new types of precursors, especially organosilicones.

Organosilicones offer several advantages. Organic groups have nearly infinite options for proper control of deposited layers. Additionally, they are compatible with other organic compounds and provide elasticity, preventing the layer to crack. Silicon atoms support binding to inorganic substrates including glass [1]. Hexamethyldisiloxane (HMDSO) and tetraethyl orthosilicate (TEOS) are used as PECVD precursors very often [2,3]. Main applications of organosilicone precursors are in microelectronics [4], automotive and food industry [5].

There are also other fields which can benefit from this type of precursors. Archaeology is one of them. Some organosilicone thin films have good barrier properties which make them suitable candidates for archaeological artefacts preservation. The aim of this work is to examine thin films based on HMDSO for possible application in archaeology. There are thousands of artefacts in museum depositories waiting for conservation. Their price is invaluable and they are being destroyed over time

Journal of Physics: Conference Series **768** (2016) 012013 doi:10.1088/1742-6596/768/1/012013

due to the lack of conservation capacity of museums. Thus there is a need for a new high capacity technique which is relatively cheap.

When conserving an artefact, several criteria have to be met:

1. Conserving layer has to provide sufficient corrosion protection.
2. Layer must not change the appearance of treated item.
3. Temperature of treatment process must not cause any changes to the item.
4. Layer must be removable by simple technological procedure.

Barrier properties are the most important parameter of all. They can be tested in corrosion chamber where salt fog and increased temperature simulate most frequent factors causing corrosion. However, the results are evaluated visuallyand cannot be quantified easily. Another approach is to measure the permeability of deposited layers. Oxygen transmission rate (OTR) measurement is a standard method for gas permeability determination [6]. In principle, it measures the amount of oxygen gone through the thin film over time. It allows us to quantify barrier properties of a thin film in few hours while single corrosion chamber test takes several days or weeks.

Second criterion demands no visible changes in appearance of the artefact. Optical changes are usually determined by colorimetry. This technique works with colour spaces using coordinates for each colour. CIE $L^*a^*b^*$ system of coordinates is currently used as the standard. Colour difference is calculated from the change of coordinates in colour space [7]:

$$\Delta E^* = \sqrt{\Delta L^{*2} + \Delta a^{*2} + \Delta b^{*2}} \qquad (1)$$

where ΔE^* is colour difference and L^*, a^* and b^* are coordinates in CIE $L^*a^*b^*$ colour space. Distinguished changes in appearance occur for $\Delta E > 0.5$.

Next, no changes can be caused thermally. Temperature may have significant influence on the artefact. PECVD is a low-temperature technique. When the setup for the deposition is correct, the temperature of the substrate does not exceed 100°C. Metallic artefacts withstand relatively high temperature without any harm. However, metallurgical changes can occur in certain metals (e.g. brass) at the temperature higher than 150°C. Non-metallic items must be treated more carefully. Material of these items can usually withstand only much lower temperatures (e.g. $t < 80$-100°C for plastic artefacts). PECVD offers an excellent option for this purpose: application of pulsed mode. Recent studies proved pulsed mode can decrease thermal stress down to about 60% in comparison with continuous discharge [8].

Fourth criterion is for PECVD the most challenging one. PECVD products can be prepared with excellent adhesion to the substrate and most researchers have been trying to find a way how to improve adhesion on their substrates. On the other hand, we need to control the adhesion quite accurately for the purposes described. The deposited layer must have sufficient adhesion to maintain idle state and protect the substrate but there must be a way how to overcome adhesion or to reduce it subsequently. We can use adhesion "demoters" to control the adhesion. Other approach is to deposit an interlayer which is compatible with both substrate (artefact) and barrier thin film and can be removed easily. This interlayer may or may not have barrier properties. Anyway it must meet the other preservation criteria. Special photoresins as well as different thin films (e.g. parylene) may meet them.

2. Experimental

PECVD was carried out in a glass bell-jar type reactor (volume of 60 L). Capacitively coupled plasma was generated by a radiofrequency source (Cesar, 13.56 MHz). The scheme of the apparatus is shown in Figure 1. Plasma discharge was formed between two disc electrodes (diameter of 14 cm). The upper electrode with substrate holder was supplied through automatic matching network by a RF source with frequency 13.56 MHz up to the power of 500 W. HMDSO was used as the precursor. Liquid precursor was transformed into vapour phase and mixed with oxygen before it entered the reactor. Oxygen and precursor flow rates (0.2 to 6 sccm) were controlled by Bronkhorst mass flow controllers to maintain total constant flow rate of 10 sccm. Whole device was pumped continuously by rotary oil pump as a primary pumping. Secondary pumping system including another rotary oil pump and a turbomolecular pump was used to clean the reactor from all impurities before each deposition. Secondary pumping was separated from the reaction chamber by a gate valve during the deposition in order to prevent the thin film deposition inside the turbomolecular pump. The precursor pressure and the total pressure in the reactor were monitored by Penning, Pirani and Capacitance gauges. Total pressure of 15 Pa remained constant during the deposition process. The optical emission spectrometer Jobin Yvon Triax 550 with a 1200 gr/mm grating and CCD detector was used for the plasma diagnostics. Optical fibre was installed at a quartz window on a long flange protected by a metallic mesh which prevents coating of the window during the deposition.

Supplied power, reaction mixture composition and duty cycle of pulsed plasma were chosen as main parameters for the deposition control. Only one parameter was varied at a time, the others remained constant. Thin films were deposited simultaneously on pure silicon substrate, iron plates, glass, polyethylenevynilacetate (PEVA, 100 μm) and polypropylene foil (PP, 20 μm). Optical emission spectra were measured during the deposition process. Oxygen transmission rate was measured for PP foils after the deposition. Infrared (IR) spectra were obtained after the deposition, as well. Thin film thickness was monitored in situ by quartz crystal monitor and verified on finished thin film by interference microscopy. Thickness of all films was 40-50 nm. It is known that thinner films do not form compact layers reliably. Thicker films crack on these substrates due to different thermal expansion coefficients.

Some substrates were cleaned by plasma before thin film deposition. Argon plasma and oxygen plasma was used for the cleaning (treatment time 15 minutes; 50 W continual regime; argon/oxygen flow rate of 10 sccm).

To decrease substrate temperature and improve barrier properties of thin films, we managed to implement duty cycle into the deposition process. In pulsed mode, both supplied power and duty cycle were varied to maintain mean power in plasma discharge (mean power 50 W, supplied power 50-500 W, duty cycle 10-100%).

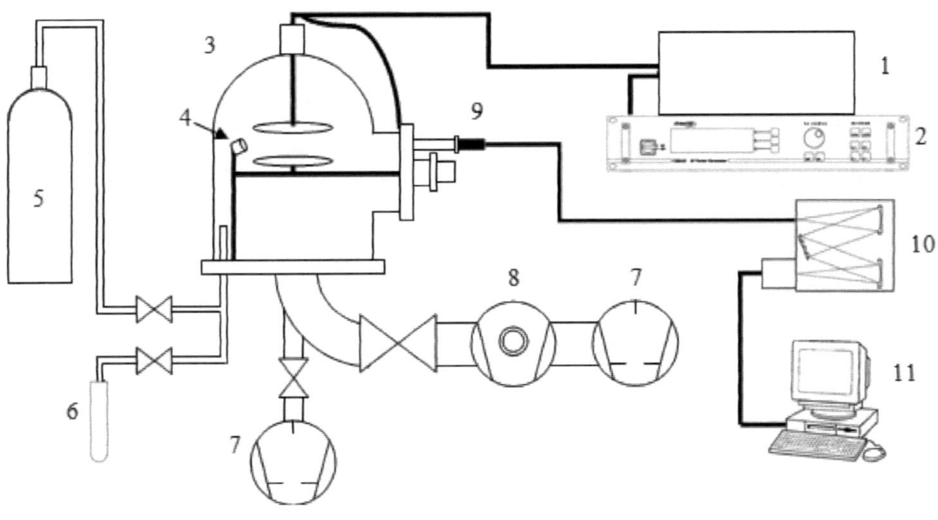

Figure 1. Apparatus for PECVD with HMDSO precursor.
1. Matching Box; 2. RF generator (13.56 MHz); 3. Deposition chamber; 4. Crystal oscillator; 5. Oxygen bottle; 6. HMDSO precursor; 7. Rotary oil pump; 8. Turbomolecular pump; 9. Optical fibre; 10. Optical emission spectrometer; 11. PC.

3. Results and Discussion

First, barrier properties of deposited thin films were tested. Gas permeability of organosilicone-based thin films is highly dependent on deposition parameters. This was confirmed by preliminary experiments. We prepared thin films with very different barrier properties. Oxygen transmission rate varied from 120 to 1500 cm^3 m^{-2} atm^{-1} day^{-1}. We focused on films with low OTR only, optimizing the parameters of the experiment to obtain as low OTR as possible. Dependence of oxygen transmission rate on supplied power and on composition of reaction mixture (O_2 content in gas mixture of HMDSO+O_2) is shown in Figure 2. Both dependences show minima corresponding to the best barrier properties achieved. Several reasons led to these results:

- Low supplied power does not enable sufficient ionization and fragmentation of the precursor.
- Too high power causes excessive fragmentation. Composition of thin films was analysed by X-ray photoelectron spectroscopy and infrared spectroscopy. Both techniques indicate composition similar to SiO_2. Because Si-O bonds are required, destroying these bonds by high power leads to lower SiO_x layer formation. Additionally, defects in the thin film occur.
- Infrared spectra (Figure 3) show decreasing carbon content in the thin film with increasing O_2 content in reaction mixture. This also corresponds to the decrease of CH fragment concentration in optical emission spectra. All that means less and less carbon incorporated in deposited thin film. The lower the carbon content, the better barrier properties can be achieved. However, certain amount of carbon is needed. When the carbon content in the thin film was too low, thin films lost their elasticity and started to crack. Incorporated carbon provides elasticity which is vital for deposition on flexible substrates.

Implementation of pulsed mode into the deposition process brought new interesting results. Dependence of OTR on duty cycle (Figure 4) showed a minimum between 20% and 30%. High power in a pulse created particles with high kinetic energy responsible for more intensive bombardment of the substrate surface. Bombardment removes solitary particles jutting out over the created thin film. It makes thin film denser with better barrier properties. Bombardment is influenced by bias, too. Bias (or self-bias, in our case) enhances particle bombardment and therefore improves barrier properties of thin film. Significant dependence of self-bias on total pressure was observed. Standard working pressure for our apparatus is about 15 Pa. When increasing the pressure over 30 Pa, the electrode with substrate holder ceases to bias self.

Barrier properties of deposited films depend on surface morphology, as well. The smoother the surface, the better barrier properties of the thin film. Several tests were run to explore the option of cleaning of the substrate by argon and oxygen plasma before thin film deposition. Argon pre-treatment decreased OTR of deposited film from 120 cm^3 m^{-2} atm^{-1} day^{-1} to 80 cm^3 m^{-2} atm^{-1} day^{-1}.Oxygen pre-treatment caused OTR to increase to about 160 cm^3 m^{-2} atm^{-1} day^{-1}. Oxygen plasma probably grafted reactive species (e.g. radicals of atomic oxygen, O_2, OH) to the surface of the substrate instead of cleaning it by simple particle bombardment. Oxygen plasma did not improve barrier properties; argon plasma pre-treatment is suitable for deposition of better barrier layers.

During preparation of thicker thin films, it is necessary to pay attention to thin film appearance and transparency. Thick SiO_x films (film thickness more than 10 μm) have white or light yellow colour and their transparency decreases. Certain thicknesses cause light diffraction. These cases must be avoided. To evaluate the change of artefact's optical appearance, colorimetry was used. Colour difference (ΔE) did not exceed 0.3 for every deposited thin film. Higher values of ΔE are expected for thicker films. Finding an optimum for film thickness and ΔE will be the subject of further research.

Temperature of the substrate was monitored during the deposition. It did not exceed 100°C in all cases. No change was observed for metallic and glass substrates. Substrates made of polymers can deform depending on polymer type. We tested PEVA and PP substrates. PEVA substrates were corrugated during depositions lasting more than 5 minutes. The temperature of these substrates was close to their melting point (105°C) causing the deformation. PP substrates (melting point 120°C) are more suitable for depositions and OTR measurement. Other types of substrates (e.g. polyethylene, ceramics) will be tested soon.

Journal of Physics: Conference Series **768** (2016) 012013 doi:10.1088/1742-6596/768/1/012013

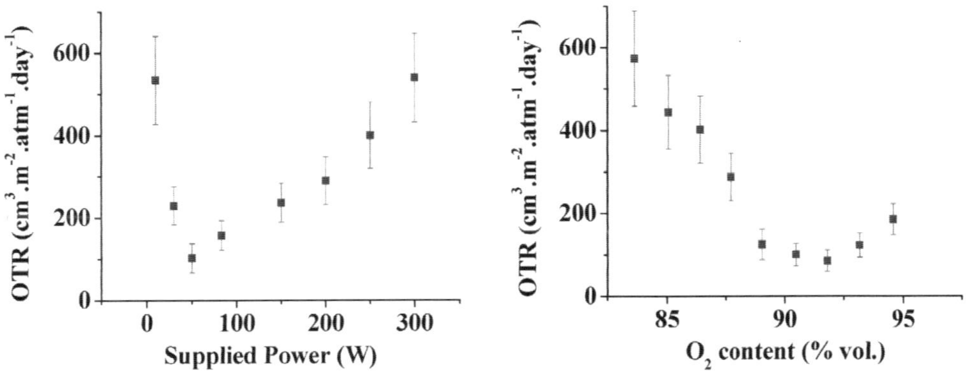

Figure 2. Dependence of oxygen transmission rate on supplied power (left) and O_2 content in reaction mixture (HMDSO+O_2) (right).

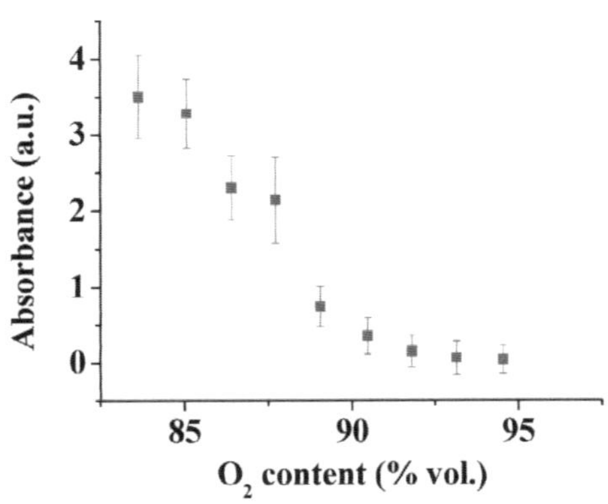

Figure 3. Dependence of IR absorbance of Si-$(CH_3)_x$ on O_2 content in reaction mixture (HMDSO+O_2).

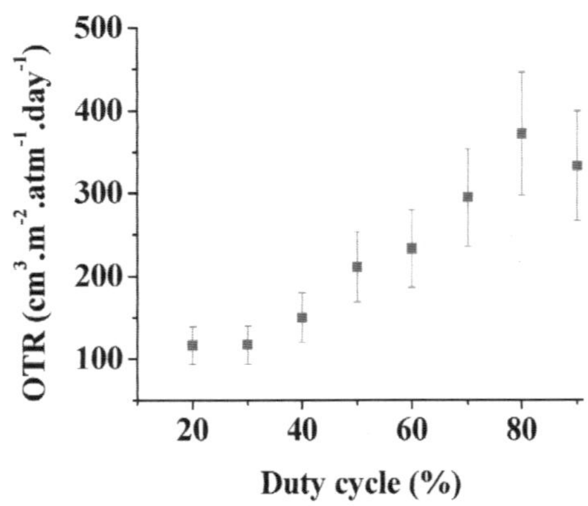

Figure 4. Dependence of oxygen transmission rate on duty cycle (mean power 50 W).

Finally, removability of thin films was tested. SiO_x layers have excellent chemical resistance so all attempts to remove these films chemically failed. Mechanical removal is quite complicate, as well. SiO_x films are very thin (40-50 nm) and have very good adhesion to the substrate surface. It is difficult to scratch them off without causing any harm to the substrate. Best solution is to use an intermediate layer which can be removed easily and that is compatible with both substrate and SiO_x thin films.

4. Conclusion

SiO_2-like thin films were tested for possible application in archaeology artefacts conservation. Thin films were deposited via low pressure PECVD using HMDSO as a precursor. They were examined whether they meet the criteria for proper conservation of artefacts.

Deposited films have good barrier properties. Best barrier properties were achieved for 50 W mean supplied power, pulsed mode of plasma with 20-30% duty cycle and 92% of O_2 in reaction mixture (O_2+HMDSO). Oxygen transmission rate of these films was approximately 120 cm^3 m^{-2} atm^{-1} day^{-1}. Argon plasma pre-treatment before the thin film deposition decreased oxygen transmission rate of thin film to 80 cm^3 m^{-2} atm^{-1} day^{-1}. Thin films have very low carbon content determined by infrared spectrometry. Colorimetric measurements proved that colour difference caused by deposition of a thin film on an item is very low. Prepared thin films do not change the appearance of the substrate.

SiO_x layer removal is complicated due to low thickness, high adhesion and excellent chemical resistance. This will be solved by an interlayer which will be removable easily. Several types are tested. Further research will focus on multi-layer systems for cultural heritage items protection as well as on temperature influence on substrates made of different materials.

Acknowledgement
This research has been supported by the Ministry of Culture of the Czech Republic, project no. DF11P01OVV004.

References

[1] Cech V, Knob A, Hosein H, Babik A and Drzal L 2012 Proc. Int. Conf. Composite Interfaces (Interface 21), 1
[2] Barni R, Zanini S and Riccardi C 2012 *Adv. Phys. Chem.* DOI:10.1155/2012/205380
[3] Hammelmann F, Heinzmann U, Szekeres U, Kirov N and Nikolova T 2005 *J. Optoelectron. Adv. M.* 7 389
[4] Lin C H, Chen H L and Wang L A 2001 *Microelectron. Eng.* **57-58** 555
[5] Morent R, De Geyter N, Van Vlierberghe S, Dubruel P, Leys C, Gengembre L, Schacht E and Payen E 2009 *Prog. Org. Coat.* **64** 304
[6] ASTM Standard F1927 – 07, DOI: 10.1520/F1927-07
[7] Ohta N and Robertson A 2006 Colorimetry: Fundamentals and Applications (John Wiley and Sons, Chichester)
[8] Sazavska V, Krcma F, Simkova T and Zemanek N 2010 *J. Phys. Conf. Ser.* **207** 12011

Determination of spatially-resolved electron temperature in nanosecond pulsed longitudinal discharges

K A Temelkov[1], I P Iliev[2], S G Gocheva-Ilieva[3], S I Slaveeva[1] and N K Vuchkov[1]

[1]Metal Vapour Lasers Laboratory, Institute of Solid State Physics, Bulgarian Academy of Sciences, 72 Tzarigradsko Chaussee, 1784 Sofia, Bulgaria
[2]Department of Physics, Technical University Sofia, branch Plovdiv, 25 Tzanko Djusstabanov, Plovdiv 4000, Bulgaria
[3]Department of Applied Mathematics and Modeling, Faculty of Mathematics and Informatics, University of Plovdiv, 24 Tzar Assen, Plovdiv 4000, Bulgaria

E-mail: temelkov@issp.bas.bg

Abstract. Using the results for the time- and spatially-averaged electron temperature, which were previously obtained by the line-ratio method of optical emission spectroscopy for several metal halide vapour lasers excited in nanosecond pulsed longitudinal discharge and analytically solving steady-state heat conduction equation for electrons as well, radial distribution of electron temperature, i.e. spatially-resolved electron temperature, is obtained.

Keywords: gas discharge, heat conduction, gas temperature, electron temperature, optical emission spectroscopy.

1. Introduction

One of the main fundamental and relevant problems in plasma physics is the determination of main plasma parameters, namely gas and electron temperatures, electron density, etc. It is very important for gas-discharge laser physics, gaseous discharges, plasma technologies, gas-discharge mass spectroscopy, absorption and emission spectroscopy and plasma in general. Electron temperature determines thoroughly the characteristics constants for elastic and inelastic electron-atom and electron-ion collisions, as well as three-body electron-ion recombination [1-4]. In particular for gas discharge lasers the electron temperature directly influences on the formation of the inversion population with the abovementioned processes or indirectly through heavy particles' interaction, such as asymmetric charge transfer, Penning ionization, etc. Widely applied methods for the electron temperature measurement include the use of a Langmuir probe and laser Thomson scattering, which are inapplicable to nanosecond pulsed longitudinal discharges (NPLDs) used in our experiments. In the literature there are several models, which predict, among the other parameters, values of electron temperature with considerable variation, and furthermore there is no overlap [1-6].

Under conditions of Local Thermodynamic Equilibrium (LTE) measurement of relative intensities of some He and Ne spectral lines, originating from different upper levels, enabled us to experimentally determine the electron temperature, which was spatially- and time-averaged in a NPLD in He, Ne and Ne-He mixtures, because spatial- and time-averaged relative intensities were measured [7-9]. Though the experimentally determined electron temperature was averaged over the time, our results, which were well accepted in gas-discharge plasma society, established line-ratio method in our research activity as a powerful source of knowledge, confirmed its applicability for NPLDs and suggested its further development and refinement. Unfortunately, time-resolved spectral investigation through devices equipped with photomultiplier are prohibited for the discharge period of NPLDs, because of the noise from high-power electrical pulses and the steep rise of voltage and current pulses of $TV.s^{-1}$ and $GA.s^{-1}$. For the discharge period, i. e. electrical power pulse period, spectral signal and noise were definitely equal, i. e. the signal remained the same using shutter in front of the spectrometer slit. That is why time-resolved electron temperature was determined in NPLD for the afterglow period only [9]. Average values of the electron temperature were also found for the afterglow period by mathematically averaging the time-resolved electron temperature over the time [9].

Unfortunately, despite the tremendous efforts, determination of radial electron temperature distribution, i.e. spatially-resolved electron temperature, encounters some insuperable obstacles for NPLDs because of the low signal and unavoidable excessive noise from the photomultiplier, i.e. low signal-noise ratio. Assuming that the electron temperature varies only in the radial direction and using the experimental results obtained for the time- and spatially-averaged electron temperature, a simple method based on analytical solution of the steady-state heat conduction equation for electrons is presented in this paper for uniform and non-uniform power input.

2. Experimental setup

Schematic diagrams of the two used discharge tubes, which are typical for the deep ultraviolet (DUV) Cu^+ Ne-CuBr laser and the middle infrared (MIR) He-(Ne)-$SrBr_2$ laser, are shown in figure 1 (a) and (b), respectively. The first basic tube (a) with an 11.8-mm inside diameter and a 15-mm outside diameter is made of fused quartz. A ceramic insert, confining the active zone, with an inside diameter of 7.1 mm, an 11.2-mm outside diameter and a length of 100 cm is coaxially sleeved in the basic tube. The CuBr powder is placed in seven quartz side-arm reservoirs, which are externally heated.

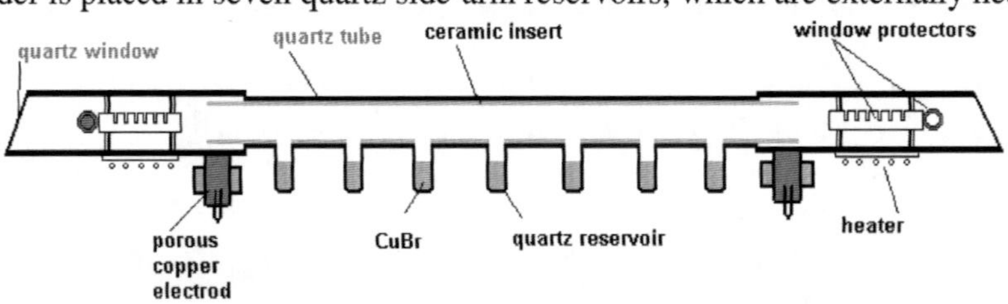

(a)

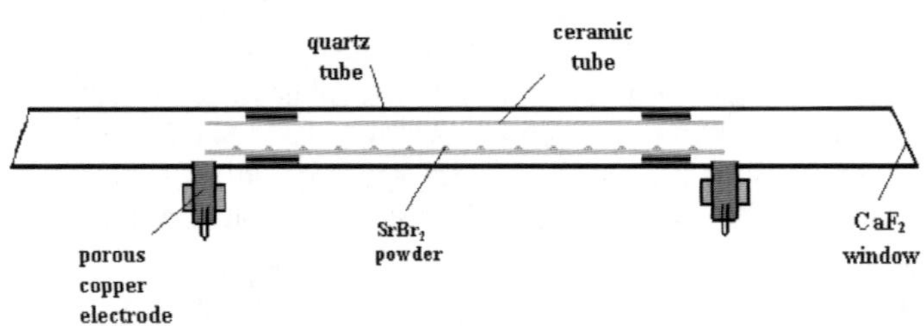

(b)

Figure 1 (a) and (b). Construction of the two gas-discharge tubes studied.

The reservoir and quartz tube temperature is measured by thermocouples. The second basic tube (b) with a 71.5 mm inside diameter and a 76 mm outside diameter is made of fused quartz. A ceramic insert, confining the active zone, with a 30.5 mm inside diameter, a 38.5 mm outside diameter and a length of 98 cm is coaxially sleeved in the basic tube. The $SrBr_2$ powder is placed inside the ceramic tube along its length. The necessary vapour pressure for laser oscillation is obtained by discharge heating; i.e. the discharge operates in a self-heating regime. The discharge tube is wrapped by a layer of fibrous insulation and its thickness depends on the power input in the discharge. The temperature at the quartz tube surface is also measured by a thermocouple. The electrodes of both tubes are made of porous copper with a special design preventing them from contamination by CuBr, $SrBr_2$ and Br_2.

The investigated NPLDs discharges are excited by an electrical pulsed scheme with interacting circuits (IC scheme). The IC excitation of the CuBr lasers, operating on the copper atom and ion transitions, was described in details in [3, 10]. The switch T is a hydrogen thyratron TGI 1000/25. The capacitors ($C_1,\ldots,C_4$), forming a capacitor bank (CB) are of the low-inductance KVI-3 type. Their capacitance is experimentally optimized for achieving of maximal output laser parameters. The capacitor bank is charged from a high voltage (HV) rectifier. The pulsed excitation scheme could deliver voltage and current pulses with an amplitude up to 25 kV and 500 A, respectively, an excitation pulse duration of 50–300 ns, and a pulse repetition frequency of 5–30 kHz. Varying the four capacitors, as well as the matching inductor L, the amplitude, waveform and duration of the pulses of the discharge current and the tube voltage, as well as the phase shift between them, could be manageably changed and controlled. In this way, appropriate discharge conditions for different excitation processes, such as electron impact excitation, charge transfer, Penning ionization, and etc., were found and hence output parameters of various metal atom and ion lasers were increased from 35 % up to two times in comparison with the frequently used electrical pulsed excitation schemes [3, 10].

The spectral investigation is made, using a Bentham M300 spectrometer equipped with two 1800 groove/mm holographic gratings and a DH-2 photo-multiplier. The spontaneous emission pulses are displayed on a Tektronix 2455A oscilloscope with a 20 MHz cutoff filter.

3. Experimental and theoretical results

3.1. Experimental determination of electron temperature spatial- and time-averaged

Under LTE conditions a measurement of the averaged relative intensity ratio of two lines, originating from different upper levels *i* and *j*, enabled us to experimentally determine the electron temperature T_e [7-9] in NPLD in He, Ne and Ne-He mixtures, using the following relative intensity ratio, Bolzmann relation for the relative equilibrium populations of the two upper levels and the expression derived from them:

$$\frac{I_i}{I_j}=\frac{\lambda_j n_i A_i}{\lambda_i n_j A_j} \qquad \frac{n_i}{n_j}=\frac{g_i}{g_j}e^{-\frac{E_i-E_j}{k_b T_e}} \qquad T_e=\frac{\left(E_j-E_i\right)}{k_b}.\left[\ln\left(\frac{I_i\lambda_i g_j A_j}{I_j\lambda_j g_i A_i}\right)\right]^{-1} \tag{1}$$

where λ_i and λ_j are the wavelength of the measured transitions from levels *i* and *j*, A_i and A_j are the transition probabilities (spontaneous rate coefficients) for the measured transitions from levels *i* and *j*, n_i and n_j are the populations in levels *i* and *j*, g_i and g_j are the statistical weights (degeneracies of levels *i* and *j*), and E_i and E_j are the energies of levels *i* and *j*, k_b is the Boltzmann constant. The use of Boltzmann's law was also justified in our previous studies [7-9]. Discharge conditions, such as discharge tube with construction and sizes described above, buffer gas or buffer-gas mixture with the corresponding pressure, and T_e^{exp} obtained experimentally, are summarized in table 1 [7-9].

3.2. Theoretical determination of electron temperature spatial-resolved and time-averaged

As is mentioned above, experimental determination of radial distribution of electron temperature with measurement of relative intensity ratio of two spectral lines for NPLDs is not possible due to some insuperable obstacles encountered, namely low signal and unavoidable noise from high-power electrical pulses in the kHz range, secondary electrical pulses in the MHz region and rise time of voltage and current pulses of $TV.s^{-1}$ and $GA.s^{-1}$. In order to obtain electron temperature distribution in the discharge zone, the following steady-state heat conduction equation:

Table 1. Discharge conditions used as input data.

Discharge tube	Buffer gas or buffer-gas mixture	Pressure (Torr)	T_e^{exp} (eV)
First discharge tube	He	10	0.59
First discharge tube	Ne	16.7	0.48
Second discharge tube	He	45	0.49
Second discharge tube	Ne-He	2.5-42.5	0.80
Second discharge tube	Ne-He	5-40	0.68
Second discharge tube	Ne-He	10-35	0.61
Second discharge tube	Ne-He	15-30	0.78

$$\text{div.}\left(k_e \text{grad} T_e\right) + \xi q_v = 0 \tag{2}$$

for electrons is firstly solved for uniform power input, where k_e is electronic thermal conductivity of the electron gas and ξq_V is electric power density deposited for electrons heating. Using Wiedemann-Franz law for electronic thermal conductivity $\frac{k_e}{\sigma} = c.T_e$, where σ is specific electrical conductivity, $c = \frac{\pi^2}{3} \cdot \left(\frac{k_b}{e}\right)^2$ is constant quantity, k_b is the Boltzmann constant, and e is the electron charge. Assuming that σ is independent of the spatial coordinates and has a value $\bar{\sigma}$, the steady-state heat conduction equation is transformed to the given equation:

$$\text{div.}\left(T_e \text{grad} T_e\right) = -\frac{\xi q_v}{c.\bar{\sigma}} \tag{3}$$

The equation (3) is of the kind, which is typical steady-state heat conduction equation:

$$\text{div.}\left(k \ \text{grad} T\right) + Q_v = 0 \tag{4}$$

where the dependence of the thermal conductivity k has the form $k = B.T^a$. Under assumption that temperature varies only in the radial direction equation (4) assumes the following form:

$$\frac{1}{r}\frac{d}{dr}\left(rk\frac{dT}{dr}\right) + Q_v = 0 \tag{5}$$

It is well-known [7-9] that the equation (5) has the following analytical solution for gas discharge zone with radius R:

$$T(r) = \left[T_W^{a+1} + \frac{(a+1)Q_V}{4B}\left(R^2 - r^2\right)\right]^{\frac{1}{a+1}} \tag{6}$$

considering the following boundary conditions: $\frac{dT}{dr}\Big|_{r=0} = 0 \qquad T\Big|_{r=R} = T_w$

For the case of equation (3), it is obvious that $B = 1$, $a = 1$, and $T_w = 0$, i.e. the kinetic energy of electrons for $r = R$ is zero as a first approximation. In this way, the solution of (3) has the form for the discharge zone:

$$T_e(r) = \sqrt{\frac{1}{2} \cdot \frac{\xi . q_V}{c.\bar{\sigma}}} \cdot \sqrt{\left(R^2 - r^2\right)} \tag{7}$$

where the constant ξ remained to be obtained. The spatially-average electron temperature in the discharge zone is found by mathematically averaging the solution (7) in the discharge zone over the radius and is made equal to the experimentally determined electron temperature T_e^{exp} (see table 1), i.e. $\bar{T}_e = T_e^{exp}$. After mathematical processing the following radial electron temperature distribution is obtained:

$$T_e(r) = 4.\frac{T_e^{\exp}}{\pi.R}.\sqrt{(R^2 - r^2)} \quad (8)$$

Radial distributions of electron temperature in the discharge zone are shown in figure 2 for NPLD in He-CuBr and Ne-CuBr mixtures.

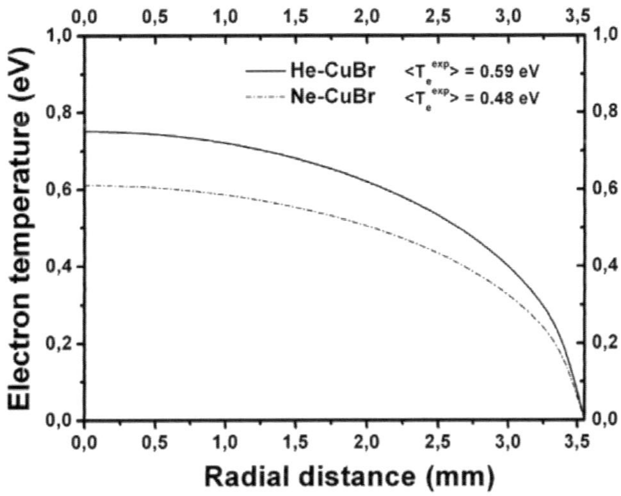

Figure 2. Radial distributions of electron temperature for NPLD in He-CuBr and Ne-CuBr mixtures.

Radial distributions of electron temperature in the discharge zone are presented in figure 3 for NPLD in pure He of 45 Torr and Ne-He mixtures with additives of neon of 5, 10 and 15 Torr maintaining the total mixture pressure of 45 Torr constant. The used average input power is also shown in the figure.

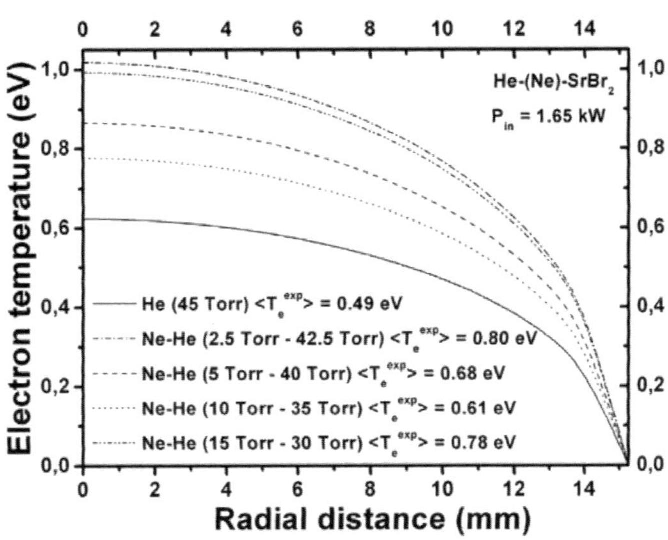

Figure 3. Radial distributions of electron temperature in NPLD in He- and Ne-He mixtures.

A comparison is made between radial electron temperature distributions for electrical power density independent of radial coordinate and electrical power density, which depends on the radius with polynomial profile of the second degree. The discrepancy is less than 10^{-5} eV and is impossible to be presented in a figure.

4. Conclusions

Using the results obtained for the spatially and time-averaged electron temperature by the line-ratio method of optical emission spectroscopy and analytically solving steady-state heat conduction equation for electrons as well, radial distribution of electron temperature, i.e. spatially-resolved electron temperature, is obtained for NPLD in various gas mixtures. The results obtained, though simplified, enable us to undertake development of numerical 2D model (*r, t*) for determination of spatially- and time-resolved electron temperature in NPLD exciting DUV Cu^+ Ne-CuBr laser and other powerful gas-discharge lasers.

References

[1] Carman R J 1990 *IEEE J. Quantum Electronics* **26(9)** 1588-608
[2] Carman R J, Brown J W and Piper J A 1994 *IEEE J. Quantum Electronics* **30(8)** 1876-94
[3] Vuchkov N K, Astadjov D N and Sabotinov N V 1994 *IEEE J. Quantum Electronics* **30(3)** 750-8
[4] Stoilov V, 1995 *MS thesis* Sofia University
[5] Badenko S M, Lakoba I S and Yakovlenko S I 1980 *Sov. Phys. Leb. Inst. Rep.* **11** 17-22
[6] Naylor G A, 1985 *PhD thesis* Oxford University
[7] Temelkov K A, Vuchkov N K, Freijo-Martin I and Ekov R P 2010 *J. Phys. D: Appl. Phys* **43** 075206
[8] Temelkov K A, Slaveeva S I and Vuchkov N K 2011 *IEEE Transactions on Plasma Science* **39(3)** 831-5
[9] Slaveeva S I, Temelkov K A and Vuchkov N K 2012 *Journal of Physics: Conference Series* **356** 012011
[10] Vuchkov N K Temelkov K A and Sabotinov N V 2005 *IEEE J. Quantum Electron.* **41(1)** 62-5

Journal of Physics: Conference Series **768** (2016) 012015 doi:10.1088/1742-6596/768/1/012015

High-rate deposition of silicon films in a magnetron discharge with liquid target

A Tumarkin[1], M Zibrov[1], G Khodachenko[1] and D Tumarkina[2]
[1]National Research Nuclear University MEPhI (Moscow Engineering Physics Institute), Kashirskoe shosse 31, 115409 Moscow, Russia
[2]Moscow State University, 119991, Moscow, Russia

E-mail: sanyahrustal@mail.ru

Abstract. Silicon coatings have been deposited on substrates made of low-carbon and high-carbon steels and tungsten in a magnetron discharge with liquid target at substrate bias voltages ranging from +100 V to -600 V. The structure of obtained coatings was examined by a scanning electron microscopy. The strong influence of substrate bias voltage on the coating structure was observed. The corrosion resistance of coated steel samples was examined in concentrated sulphuric, hydrochloric and nitric acids and their solutions. The resistance of coated tungsten samples against high-temperature oxidation was examined by their exposure to O_2 gas at a pressure of 0.2 Pa and a temperature of 1073 K. The coatings deposited under bias voltages of +100 V and -600 V had dense structures and showed the best protective properties among all deposited coatings.

1. Introduction

Silicon-based coatings have been widely used for many years in microelectronics, photonic devices, as adhesive layers for diamond-like carbon coatings, etc. In particular, siliconizing is a common method for protection of articles made of various metals (steels, titanium alloys, molybdenum, etc.) against corrosion in aggressive media and high-temperature oxidation [1-3]. However, in the latter case the coatings obtained by common methods (laser surface alloying, vapour-phase siliconizing, and powder siliconizing) usually have quite high porosity, thus they cannot provide excellent corrosion resistance. Coatings obtained by magnetron sputtering usually have a very dense structure and good adhesion to a substrate [4]. Nevertheless, the deposition rates in classical magnetron sputtering systems (MSS) are much lower than those in chemical vapour deposition (CVD) and evaporation systems. A magnetron discharge with liquid target (MDLT) is a promising tool for high-rate deposition of high-quality coatings [5]. The remarkable feature of this discharge is that due to ion bombardment of the cathode, at a certain value of the discharge power the cathode material starts to melt and intensively evaporate. Thus, the flux of atoms on a sample surface is a combination of fluxes of sputtered end evaporated atoms from a cathode. This allows to reach much higher deposition rates (in comparison with that in conventional magnetron sputtering systems), to heat a substrate to high temperatures due to radiation from a cathode and energy of arriving particles, and to have an ion assistance during the deposition (i.e. to influence on the film structure by biasing the substrate).

This work is devoted to investigation of structure and protective properties of silicon films deposited in a MDLT on different substrates at various conditions.

2. Experimental

The coatings were deposited on mirror-polished plates made of low-carbon steel (transformer steel type 3) and high-carbon steel (C1035), and on 50 μm tungsten foils in a DC magnetron discharge with liquid cathode in the PINCH installation [5]. Silicon target (with a purity of 99.9%) was located in a carbon crucible, which was thermally insulated from a water-cooled magnetic system. The samples were placed on an object table located at a distance of 18 cm above the crucible. The residual pressure of about 1×10^{-3} Pa in the chamber was obtained by a turbomolecular pump in series with a rotary vane pump. Prior to deposition, the substrates were ultrasonically cleaned in an acetone and then sputter-cleaned in a DC glow discharge in Ar initiated between the object table and walls of the vacuum chamber. At the beginning, a DC magnetron discharge in Ar at a pressure of 1 Pa with a power of 2.5 kW was ignited. Due to poor heat removal from the crucible, silicon in it started to melt and then intensively evaporate. This discharge regime is referred as MDLT. The coatings were deposited under various DC bias voltages at a sample holder (-600V, -450V, -300V, -150V, 0V, +100V).

The structure of deposited films was investigated in a scanning electron microscope (SEM) TESCAN VEGA3. Both the surface and the fracture cross-sections of the deposited films were studied. From these observations the film thickness was measured. The deposition rates were estimated by dividing the coating thickness by the deposition time.

In order to check the protective properties of deposited films on steel substrates, the samples (both coated and uncoated) were exposed to concentrated sulphuric, hydrochloric and nitric acids for 30 min. The samples that were not destroyed during these tests were then exposed to 10% aqueous solutions of sulphuric, hydrochloric and nitric acids for 24 hours. After the exposure, the samples were washed in distilled water and then their surface was investigated in an optical microscope. For testing of the protective properties of obtained coatings on tungsten samples against high-temperature oxidation, the samples were exposed to O_2 gas at a pressure of 0.2 Pa and a temperature of 1073 K for 30 min. The erosion rate of the samples was estimated by their weight loss after the exposure.

3. Results

SEM investigations demonstrated that the structures of coatings deposited in the identical conditions were almost similar for all used types of substrates. Examination of the fracture cross-sections of the samples indicated that with increasing the bias voltage the structure of the coatings became denser (**Figure 1. a–c**). Besides, the coatings deposited at +100 V bias also had a dense structure (**Figure 1. d**).

The deposition rate at the bias voltages of +100 V and 0 V was about 1.2 μm/min, and decreased with increasing the negative bias voltage (to 0.2 μm/min at -600 V) due to increase in sputtering rate of the film by arriving ions.

The corrosion tests of the bare steel samples indicated their intense erosion in all used acids. The samples coated with Si films deposited at the bias voltages of 0 V, -150 V, and -300 V corroded much slower than the bare samples, albeit the coatings slowly destructed with time and the corrosion rate of the samples gradually increased. Neither substrate corrosion nor film destruction was observed for the samples with coatings deposited at -450 V, -600 V, and +100 V after their short-term exposure in concentrated acids. After the long-term exposure of these samples in solutions of these acids, the coatings deposited at -450 V partially destructed (some parts of the coatings exfoliated), whereas on the coatings deposited at -600 V and +100 V only small pinholes with a diameter of less than 50 μm were observed.

Similarly, the tests on behaviour of tungsten coated with the Si films at high temperatures in oxygen environment revealed that only the coatings deposited at -600 V and +100 V bias allowed to reduce the erosion rate of tungsten by more than one order of magnitude and the coatings almost did not destruct during the tests. The coatings deposited at other bias voltages showed much smaller reduction of the erosion rate of the tungsten substrate due to exfoliations of large parts of coatings during the tests.

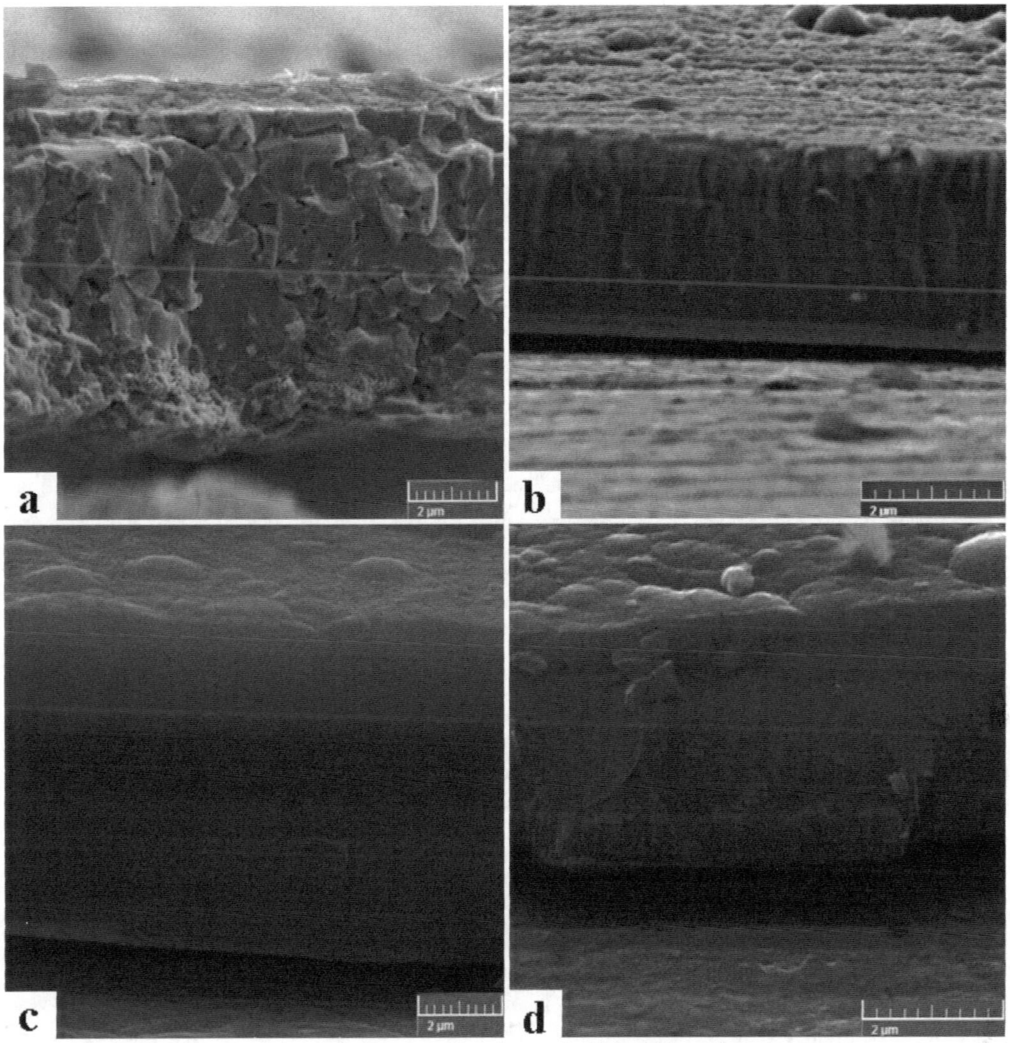

Figure 1. SEM images of the fracture cross-sections of silicon films deposited on low-carbon steel plates at various bias voltages: (a) 0 V; (b) -300 V; (c) -600 V; (d) +100 V.

4. Discussion

The destruction of the coatings during the tests may be caused by the presence of large number of through channels (for acid transport to the substrate) in the coatings and also may be due to poor adhesion of the coatings to the substrate.

The results clearly demonstrated the strong influence of substrate bias on the microstructure and, possibly, the adhesion of silicon coatings deposited in a MDLT, despite the fact that the number of ions arriving on the substrate is much less than the number of neutral particles (sputtered and evaporated). The fact that the coatings deposited at the bias voltage of -600 V had the densest structure and the best adhesion to the substrate among all coatings deposited at negative bias voltages (taking into account that the thicknesses of all of them were similar) may be due to high energy of ions arriving on the substrate resulting in enhancement of surface mobility of coating atoms, and, consequently, to structural rearrangements in the growing film [6]. Besides, ion bombardment results in increase of the coating temperature. Superior properties of the films deposited at +100 V may be explained by the fact that electron bombardment of the film leads to increase of its temperature, which was confirmed by visual observations – the sample glowed red during the deposition. One should mention that such effect was not observed during the deposition of films under negative bias voltages due to the fact that the electron current in the former case was much higher than the ion current in the latter case. It is known that elevated substrate temperature during the deposition leads to structural changes in a growing film [6]. However, other effects of inelastic interactions of electrons with the film, such as activation of adsorption centres on its surface, may also contribute.

Increase in substrate temperature may also lead to enhancement of diffusion of silicon atoms into the substrate that may result in better adhesion of the coatings to the substrate due to formation of the transition layer. In the case of ion irradiation, ion-induced mixing of coating atoms and substrate atoms may also lead to formation of such layer.

4. Conclusions

Silicon coatings were deposited on steel and tungsten substrates in a magnetron discharge with liquid target at substrate bias voltages ranging from +100 V to -600 V. It was shown that the bias voltage (both negative and positive) strongly affects the structure of deposited coatings. The coatings deposited at the bias voltages of -600 V and +100 V had a dense structure and demonstrated the best protective properties against both the corrosion in acids and high-temperature oxidation. The maximum achieved deposition rate of 1.2 μm/min (at +100 V bias) was much higher than that obtained in a conventional magnetron discharge with solid target [7].

Acknowledgements

Helpful discussions with A. Kaziev and A. Pisarev are gratefully acknowledged. This work was supported by the Russian Ministry of Education and Science (Contract 14.Y26.31.0008).

The results of this work were produced in the joint project of USATU (Ufa State Aviation Technical University) and UMPO (Ufa Engine Industrial Association) with title "Elaboration and industrial development of high-precision shaping coordinated technologies and superficial hardening of responsible details from Al-alloys with heightened constructional energy efficiency", sponsored by the Resolution of the Russian Federation Government No 218 from April 9, 2010.

References

[1] Fukumoto M, Tachikawame C, Matsuzaka Y, et al 2012 *Corrosion Science* **56** 105-113
[2] Henderson A S, Slowter E E and Gonser B W US patent 2501051 A.
[3] Huang H L, Lee T Y and Gan D 2006 *Materials Science and Engineering:A* **422** 259-265
[4] Depla D, Mahieu S and Greene J E 2010 *Handbook of Deposition Technologies for Films and Coatings (*Boston: William Andrew Publishing*)* ed P. M. Martin Chapter 5 pp 253-296.
[5] Zibrov M S, Khodachenko G V, Tumarkin A V, et al. 2013 *Journal of Surface Investigation* **7** 1156-1162
[6] Reichelt K and Jiang X 1990 *Thin Solid Films* **191** 91-126
[7] Boo J-H, Park H K, Nam K H, et al. 2000 *Surface and Coatings Technology* **131** 211-215

Ion flux and energy virtual sensor for measuring ion flux and energy distribution at a RF biased electrode in ICP reactor (RIE-mode)

M A Bogdanova[1,2], D V Lopaev[1] and S M Zyryanov[1,2]

[1] Lomonosov Moscow State University, Skobeltsyn Institute of Nuclear Physics, Leninskie gory, GSP-1, Moscow 119991, Russian Federation
[2] Lomonosov Moscow State University, Faculty of Physics, Leninskie Gory, Moscow 119991, Russian Federation

E-mail: bogdanova.marya@mail.ru

Abstract. The modern technology of micro- and nanoelectronics involves a great number of steps, such as pattern transfer, where Reactive Ion Etching (RIE) in rf plasma reactors is widely used. To control the etching process, the ion flux and ion energy distribution should be managed precisely. However, the measurements of these parameters during the process in the real-time operation are impossible. This paper is devoted to the construction of a virtual diagnostics of the Ion Energy Distribution (IED) function. This method for the determination of the ion energy spectrum on the surface of rf biased electrode is based on model calculations using in-situ measured discharge parameters. The results of IED virtual diagnostics were compared with data, obtained by Retarded Field Energy Analyzer (RFEA). This was done for Ar- and H_2-plasmas operated under low-pressure rf plasma conditions. The good agreement between the model and the experimental justifies the conclusion that the IED virtual diagnostics can be applied successfully. This enables the in-situ monitoring of the IED at the electrode surface in RIE reactors.

1. Introduction

Pattern creation for the production of microstructures cannot be achieved without Reactive Ion Etching (RIE). This RIE process is carried out by placing samples on the surface of an rf biased electrode, as rule in an asymmetric rf low-pressure discharge. To control the etching process precisely it is essential to follow the ion flux and the ion energy distribution (IED) at the surface of the rf biased electrode. However, both the IED and ion flux cannot be measured directly in an industrial plasma reactor during the etching process. Therefore the development of some virtual plasma diagnostics for IED measurements is of great interest. In other words, such diagnostics is supposed to predict the energy distribution and the flux of the ions bombarding the sample, by using model calculations. The model input is formed by the set of easy measurable discharge parameters; the model output, being the ion flux and energy distribution, will be shown in the real-time on a computer screen. In the simple case of a low-pressure plasma the IED calculations can be based on a straightforward physical model of the ion motion in a collisionless rf sheath. Moreover, the modeling should be rather fast to be fulfilled in the real-time operation mode.

2. The model of the ion motion in an RF collisionless sheath

The simplest approach to describe the ion energy spectrum at a rf-biased electrode is based on the model of collisionless rf sheath [1]. Such an approach allows obtaining the IED at a CCP rf biased surface by using simple physical relations. Since the response of the ions to rapid changes of a rf sheath potential is negligible, we may introduce the notion of the mean ion energy $e\overline{V_s}$ gained in the sheath, where $\overline{V_s}$ is a mean sheath potential. This potential can be calculated, using two measurable external discharge parameters: namely the amplitude of the rf voltage $\overline{V_{rf}}$ applied to the electrode and the negative dc self-bias voltage V_{dc}, The latter originates from the difference between the areas of the biased and grounded electrodes, i. e. from the asymmetry of the rf discharge. Thus we get:

$$\overline{V_s} = \frac{1}{2}\left(\overline{V_{rf}} - V_{dc}\right) \tag{1}$$

In the case of the collisionless rf sheath the ion energy spectrum $F(E)$ can be calculated by the following way [2]:

$$F(E) = \frac{2\Gamma_i}{\Delta E}\left[1 - 4\left(\frac{E - e\overline{V_s}}{\Delta E}\right)^2\right]^{-1/2}, \tag{3}$$

where Γ_i – is ion flux to the rf biased electrode while the ion energy

$$E = e\overline{V_s} \pm \Delta E, \tag{2}$$

depends on the ion energy addition ΔE – being caused by the influence of the rf voltage applied to the electrode on the ion motion in the sheath.. The magnitude of the energy addition ΔE is determined by the ion response to applied rf voltage and is a function of the ratio between the ion pass time through the sheath τ_i and the rf period τ_{rf} :

$$\Delta E = 2e\overline{V_{rf}}\left[1 + \left(\frac{2\pi}{3}\frac{\tau_i}{\tau_{rf}}\right)^2\right]^{-1/2}, \tag{4}$$

$$\tau_i = 3s\sqrt{\frac{M}{2e\overline{V_S}}}, \qquad \tau_{rf} = \frac{1}{f}, \tag{5}$$

Here M – is the ion mass and f – the frequency of applied rf voltage. The sheath thickness s can be obtained from the well known Child law, which modified for the case of rf sheath reads [2]:

$$s = \left(0.82\frac{\varepsilon_0}{\overline{J_i}}\right)^{1/2}\left(\frac{2e}{M}\right)^{1/4}\overline{V_s}^{-3/4}, \tag{6}$$

With ε_0 – the dielectric constant and $\overline{J_i}$ – the mean ion current density on the rf biased electrode.

3. Experimental setup

The described model for the ion energy spectrum was used to develop the IED virtual diagnostics. The experimental set up is shown in fig.1. Experiments in Ar- and H_2-plasma at pressure < 20 mTorr were carried out in order to verify the proposed approach for plasmas with different ion mass and ion composition. To generate plasma an ICP rf coil was used operated at 13.56 MHz. The frequency of the CCP rf bias, applied to the electrode, could be varied from 1 till 27 MHz. The plasma density n_e and electron temperature T_e were measured by an rf-compensated Langmuir probe. An rf-compensated Retarded Field Energy Analyzer (RFEA) was used for ion energy spectrum measurements. In addition the mass-spectrometer was attached to the plasma chamber in order to obtain ion composition (figure 1).

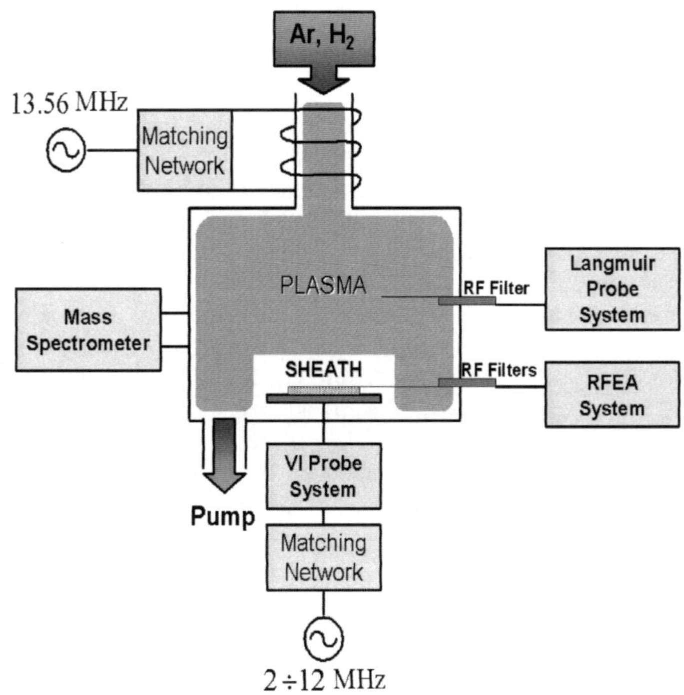

Figure 1. Experimental setup.

4. Results and discussion

4.1. Ion composition

To perform the calculations of the ion energy spectrum it is necessary to know the ion composition of plasma. In the simple gas, such as pure Ar, it is reasonable to assume that only one sort of ion Ar^+ is present; while in H_2 the ion composition can be rather complex: H^+, H_2^+, H_3^+. To take this into account the ion composition was measured by mass spectrometry. Examples of mass spectra in Ar- and H_2-plasma are shown in figures 2a and 2b respectively.

As expected, Ar^+ is the main ion in Ar-plasma (ArH^+ density is notably lower), while in H_2-plasma there are several type of ions– H_3^+, H_2^+ and H^+. Besides there is a small density of water ions H_2O^+, which is typical for plasma research done is stainless steel chambers.

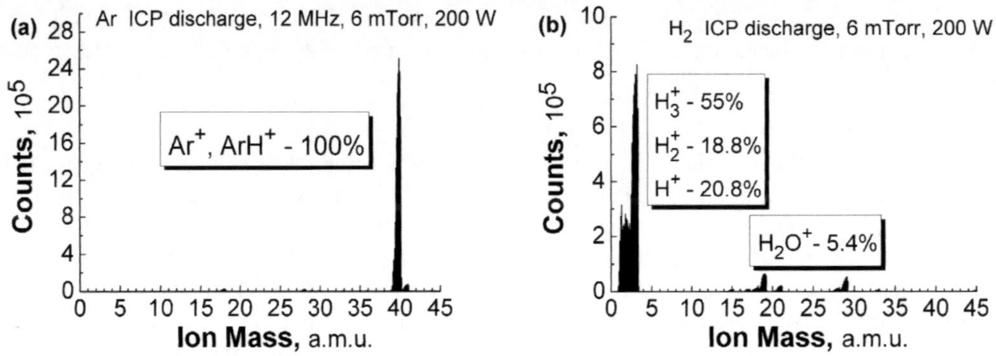

Figure 2. Ion composition: (a) – Ar-plasma; (b) – H_2-plasma.

4.2. Ion energy spectra
In the discussion on experimental data and the IED calculations it is reasonable to consider separately two cases: when the electrode is grounded and when it is rf biased.

4.2.1. Grounded electrode. To characterize the pure ICP plasma, the IED at the grounded electrode was measured. It is well known that at low pressure the Tonks-Langmuir collision -less pre-sheath model [2] can be used. The assumption of both collisionless sheath and presheath is then justified and the IED is presented as a monoenergetic peak at energy equal plasma potential.. The pre-sheath thickness is determined by the ion mean free path while ions gain the energy equal $T_e/2$ in the sheath boundary (T_e is electron temperature). Ion collisions in the presheath region lead to a widening of the peaked IED on the sheath boundary. Therefore it is quite reasonable to assume the presence of a Boltzmann distribution for ions entering the sheath with a temperature equal to $T_e/2$. Comparison with the experimental results clearly shows that the measured IED is somewhat wider than $T_e/2$. It is related to capacitive coupling with the ICP antenna that transforms part of the 13.56 MHz voltage to the sheath. The amplitude of the rf voltage at 13.56 MHz was measured by a VI probe and the influence of the capacitive coupling on the ion motion was calculated using the method described in the section 2. Examples of the calculations for Ar- and H_2-plasma are shown in figures 3a and 3b respectively.

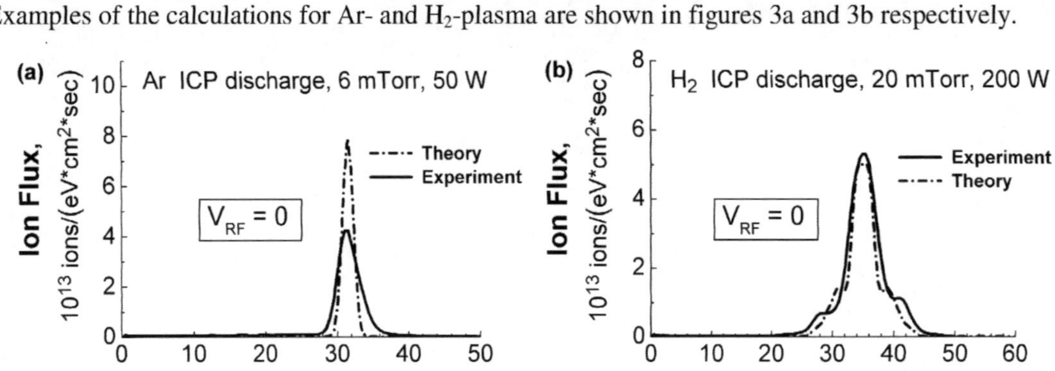

Figure 3. The IED on the grounded electrode – experimental data and theoretical calculation: (a) – Ar-plasma; (b) – H_2-plasma.

Since the amplitude of the V_{rf} at 13.56 MHz does not change by applying the rf bias to the electrode so that $V_{13.56MHz} + V_{rf\_bias}$ was really applied to the plasma sheath, we can consider the IED on the grounded electrode as "the apparatus function" for the IED calculations.

4.2.2. RF biased electrode. The experimental IEDs, obtained by the RFEA for several values of the rf bias V_{rf} at 12 MHz are shown in figure 4 for both the Ar- and H_2-plasmas. It is clear that the ion response gets more intense with increasing V_{rf} and that the IED becomes wider. Figure 4 also shows that this tendency is observed for both Ar^+ ions and for the ions generated in H_2-plasma. Moreover, it turns out that the IED-width in H_2-plasma is notably wider. This is due to the fact that lighter ions are more sensitive to the rf field.

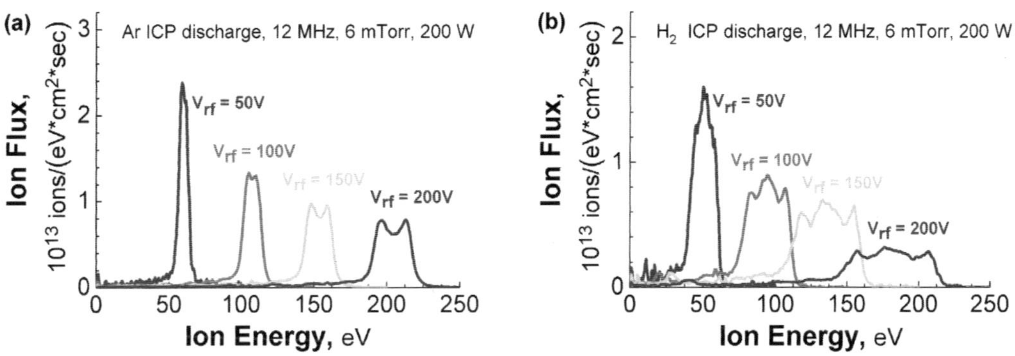

Figure 4. The experimental IEDs measured by the RFEA for different values of rf bias: (a) – Ar-plasma; (b) – H_2-plasma.

Besides, in the H_2-plasma, where ion composition is complex, the dependence of the IED shape on the ion mass is also observed: the heavier ions have a smaller response to applied rf field, and are consequently more peaked. So the narrowest small peak on the IED top at the mean ion energy ??eV_s corresponds to H_2O^+ ions, which have the biggest mass. The other two peaks, situated symmetrically relative to H_2O^+ energy peak, correspond to H_3^+, H_2^+ and a part also to H^+ ions. The calculated IEDs with the ion composition measured by mass spectrometer are shown in figure 5. It is seen, that the model calculations rather well reproduce the experimentally measured IEDs (see figure 4). There is some small difference only at lower energies connected with some rare ion collisions in the sheath.

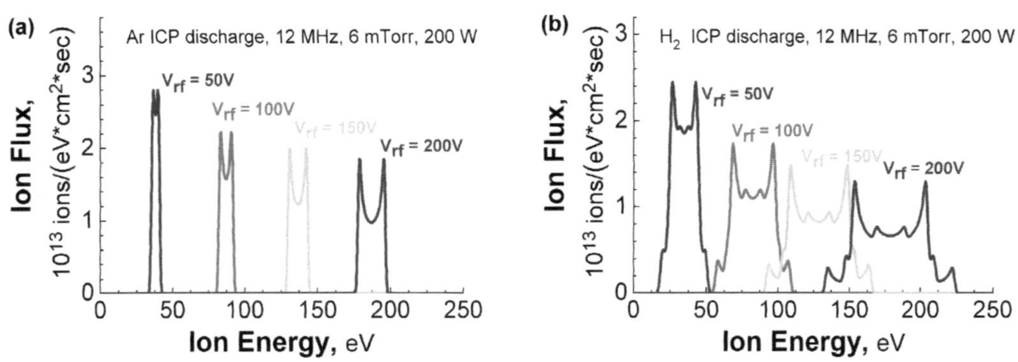

Figure 5. The model IEDs for different values of rf bias: (a) – Ar-plasma; (b) – H_2-plasma.

4.3. Ion flux

For the IED normalization it is necessary to calibrate the RFEA accurately . In principle this can be done by inserting the Langmuir probe data (n_e and T_e) into the Bohm relation. But in absence of the probe, i.e. the typical situation for industrial plasma reactors, it should be done by an different method. To normalize the IEDs on the ion flux a special method is developed by using an alternating dc self-bias; in this method the electrode acts as a "flat probe". When the rf bias is switched off fast enough, the stored charge on the capacitance associated to dc self-bias V_{dc} will be lost due to ion current to the electrode. The dependence of the current on V_{dc} (see figure 6) presents the ion part of VAC of the flat probe, i.e. electrode. From this VAC the electron temperature T_e, floating potential V_f and ion current I_{dc} (or ion flux) can be extracted.

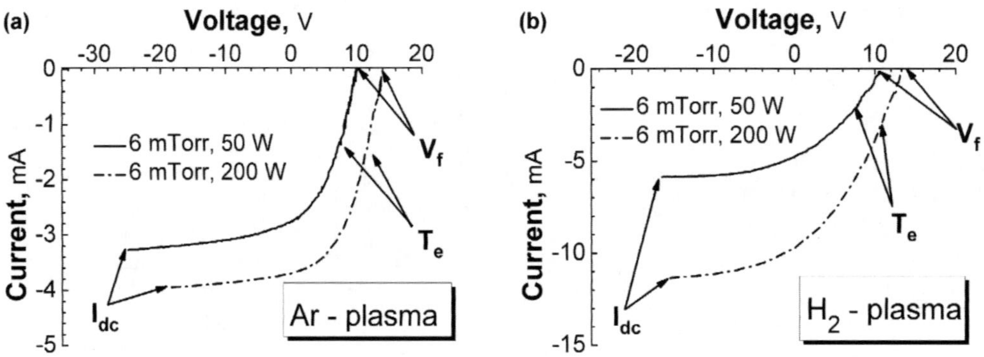

Figure 6. VAC (ion part) of the electrode that function as a "flat probe" for two power values: (a) – Ar-plasma; (b) – H_2-plasma.

5. Conclusion

The concept of "virtual diagnostics" is developed for the in-situ measurement of the ion energy distribution (IED) of ions impinging on an rf-biased electrode in a low-pressure RIE reactors. In this way the IED is found from external electric discharge parameters by using model calculations. The proposed approach was verified by experiments in an Ar- plasma and for a complex-composited H_2-plasma at pressures below 20 mTorr. In the experiments mass-spectrometer, an rf compensated Langmuir probe and an rf compensated RFEA were used to characterize the plasma and to find the IED at rf biased electrode. A simple model of collisionless rf plasma sheath was applied to calculate IEDs from measured discharge parameters and shows a good agreement with IEDs measured by RFEA. This works for both Ar- and H_2- plasmas. Thus approach of "IED virtual diagnostics" can be successfully applied to the control of the ion flux and energy in the case of a collisionless plasma sheath what is typical for conditions of low-pressure RIE reactors.

Acknowledgements

This research is supported by SRC program Contract 2012-KJ-2280, by the Grants of Russian Foundation of Basic Research (12-02-00536-a) and by 14-02-31599 mol-a and by Optec grant № 28/2014/71-Msk

References

[1] Gahan D, Daniels S, Hayden C, O'Sullivan D and Hopkins M B 2012 *Plasma Sources Sci. Technol.* **21** 015002
[2] Michael A Lieberman, Allan J Lichtenberg 2005 *Principles of plasma discharges and materials processing* (John Wiley & Sons, Inc., Hoboken, New Jersey)

Plasma reactor for deposition of carbon nanowalls at atmospheric pressure

Zh Dimitrov, D Mitev and Zh Kiss'ovski
Faculty of Physics, Sofia University "St. Kliment Ohridski", 6 James Boucher Blvd, Sofia, Bulgaria

E-mail: cozhiv@gmail.com, kissov@uni-sofia.bg

Abstract. In this study a novel plasma reactor for deposition of carbon nanowalls at atmospheric pressure is constructed and characterized. A low power microwave discharge is used as a plasma source and working gas of $Ar/H_2/CH_4$ gas mixture. The substrate is heated by plasma flame and its temperature is in the range 600-700 C. The chemical composition of the plasma and the gas mixture effect on the concentration of the various particles in the plasma is investigated by optical emission spectroscopy. The emission spectrum of the plasma jet in $Ar/H_2/CH_4$ mixture shows the presence of carbon (Swan band) and an intensive line of CH (388 nm), which are necessary species for deposition of carbon nanostructures. Additional voltage in the range from -20 V to -100 V is applied in order to ensure the vertical growth of graphene walls. Results of deposited carbon nanostructures on metal substrate are shown.

1. Introduction

Graphene and carbon nanostructures of several graphene layers draw attention with numerous applications in microelectronics [1], for sensors [2] and super capacitors [3]. In recent years research has been focused on deposition of vertically oriented graphene (carbon nanowalls) due to its specific morphology - intersecting vertical walls with sharp edges and enormous ratio of surface area to the volume. Plasma-enhanced chemical vapor deposition (PECVD) is a key technology for the preparation of carbon nanowalls. Variety of gas discharges (DC, RF and microwave) are used for deposition of nanowalls, but they operate in a vacuum chamber at low pressure, which leads to long periods of deposition. In recent years carbon nanowalls are successfully obtained in a normal glow discharge at atmospheric pressure by plasma enhanced chemical vapor deposition [4] in $Ar/H_2/CH_4$ gas mixture by addition of H_2O. The mirror-polished stainless steel anode plate is used as a substrate for deposition. Microwave plasmas are widely used for synthesis of graphene [5 and 6] and carbon nanostructures [7]. Microwave discharges are characterized with high density plasma and relatively high gas temperature, which cause quick decomposition of hydrocarbon gas and production of many reactive species.

In this study a novel plasma reactor for deposition of carbon nanostructures is presented. A low power microwave discharge is used to create plasma at atmospheric pressure in a gas mixtures of $Ar/H_2/CH_4$. The chemical composition of the plasma flame is investigated by optical emission spectroscopy and results show that the plasma in the reactor is suitable for deposition of carbon nanowalls at atmospheric pressure. Obtained plasma parameters and measurements of substrate temperature demonstrate the possibility for heating of substrate to temperatures above 600 C.

2. Experimental set-up and diagnostics

A plasma reactor is presented in figure 1. It consists of dielectric (quartz) chamber with cylindrical shape with two metal flanges (top and bottom). The miniature microwave plasma source is fed with a signal from MPG – 4M generator (0-100W; 2.45 GHz) through a coaxial line. The exciter of plasma surface waves has an outer diameter of 2 mm and 1mm thick steel capillary as inner conductor. The forward and reflected powers are measured with Pasternack PE 2219-30 directional coupler. In order to withstand the high gas temperature a thin alumina ceramic tube is used as discharge tube and as a dielectric in the coaxial exciter.

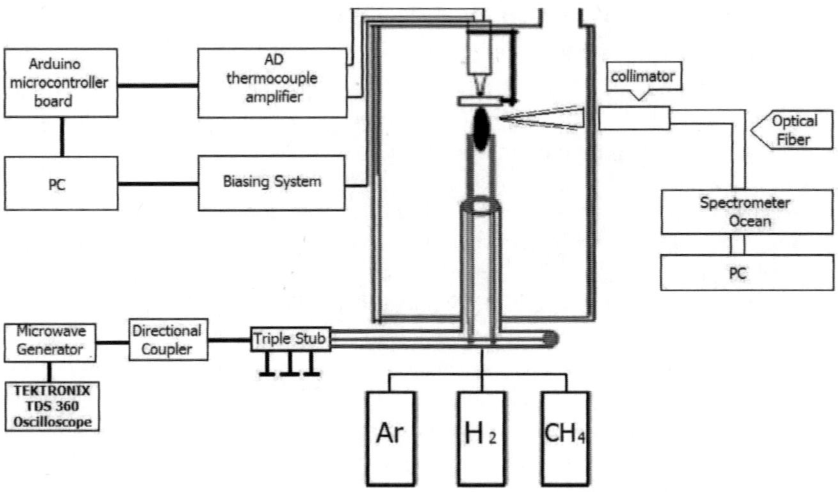

Figure 1. Experimental set-up and diagnostics

The tuning of the source at 2.45 GHz is carried out by a Maury 1878C triple stub. Indication for precise matching is the minimum of reflected power. The plasma source can self-ignite in continuous and pulse regimes at a specific threshold value of the input power. Pulses have frequency in the range of 70-700 Hz with duration of 100 μs to 14 ms. Pulse parameters are recorded with Tektronix TDS 360 oscilloscope. A gas mixture ($Ar/H_2/CH_4$) let through the capillary and gas flow ratios are controlled by APEX-AX-MC mass-flow meters. The holder of the metallic substrate for deposition is made of stainless steel tube, that includes teflon head. Heating of the substrate is controlled mainly by pulse frequency, duration and distance between the plasma source and the substrate [8]. To reach the required substrate temperature for certain compositions of nanostructures a mini heater is also used. Additional voltage in the range -20 V to -100 V is applied to the substrate to decrease the time needed for surface cleaning and to assists the vertical growth of graphene walls. The direct current due to this voltage causes further heating of the substrate. The temperature of the substrate is measured by an electronic system with a thermocouple [8].

The species formed in the active region of the plasma jet in a pulse regime are investigated with HR Ocean Optics spectrometer. The light from the plasma column positioned between the discharge tube end and metallic substrate is collected by a lens system (collimator) coupled with an optical fibre and the raw data is processed with a PC.

3. Results

The plasma reactor is constructed for deposition of carbon nanostructures on small metallic plates (1.5 mm x 1.5 mm). Ni and Cu plates are subjected to a chemical cleaning by ethanol followed by

10% hydrofluoric acid bath. After positioning the plate above the plasma source in the reactor, it is subjected to plasma cleaning in argon-hydrogen plasma at substrate temperatures below 400 C for a half hour. Experiments are carried out at frequency 100 Hz, microwave signal and pulse duration in the range 100 μs - 4 ms, which ensures gradual increase of the substrate temperature and stable plasma operation. The gas temperature, obtained via simulation of OH (306-310nm) band, reaches 2000 K at the tube end. In the process of understanding the deposition of carbon nanostructures, optical spectroscopic measurements were performed of two different gas mixtures - an Ar /H_2 (240 sccm/10 sccm) and Ar/H_2/CH_4 (239 sccm/10 sccm /1 sccm). The chemical decomposition of methane in the plasma and the influence on the appearance of the various reactive species of the gas volume ratios in the mixture are investigated using the emission spectra of the plasma jet.

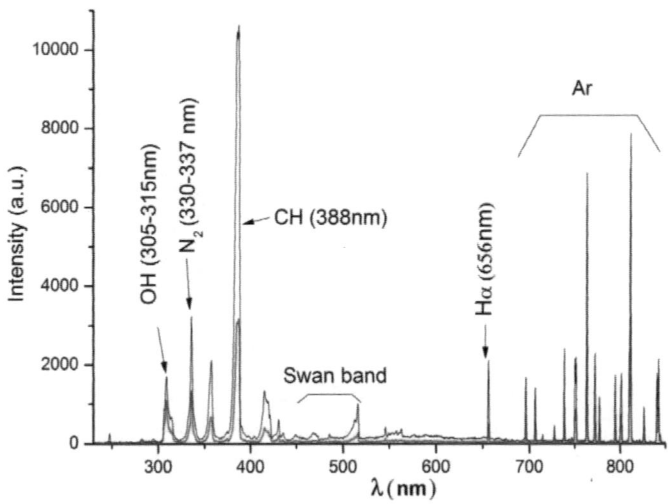

Figure 2. Emission spectra of plasma flame in Ar/H_2 (red) and Ar/H_2/CH_4 (black) gas mixtures

Argon spectral lines are recorded at the wavelength range from 690 to 900 nm under different experimental conditions explored in the study. An example of spectrogram in the first gas mixture (Ar/H_2) is shown in figure 2 (red line). Atomic argon lines in the red/near-infrared region are clearly dominating the spectra due to the contribution of the radiative transitions between the excited levels of the $3p^5 4s$ and $3p^5 4p$ configurations. Strong H_α line and OH band are also observed in the spectrum, while the intensity of N_2 band decreases with the pretreatment time.

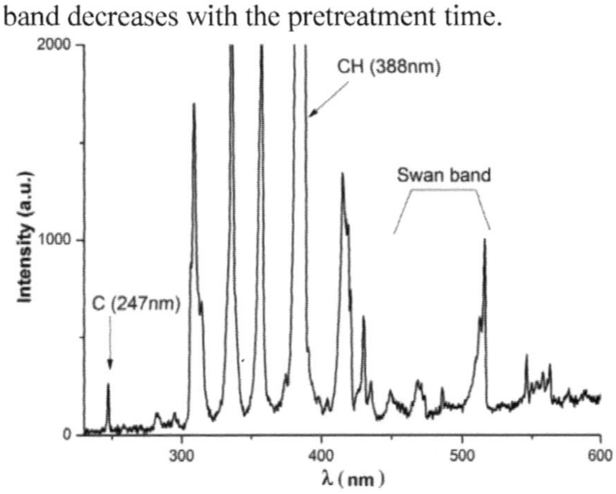

Figure 3. Spectrum of Ar/H_2/CH_4 plasma

At substrate temperature above 600 C in the $Ar/H_2/CH_4$ plasma a recorded spectrogram (figure 2) is used for identification of excited species. Due to decomposition of methane additional lines are observed (Figure 3). Spectrum shows presence of atomic carbon line at 247.8 nm, C_2 band system (Swan band) and an intensive CH (388nm) line. The presence of these species is a prerequisite for carbon nanostructures deposition.

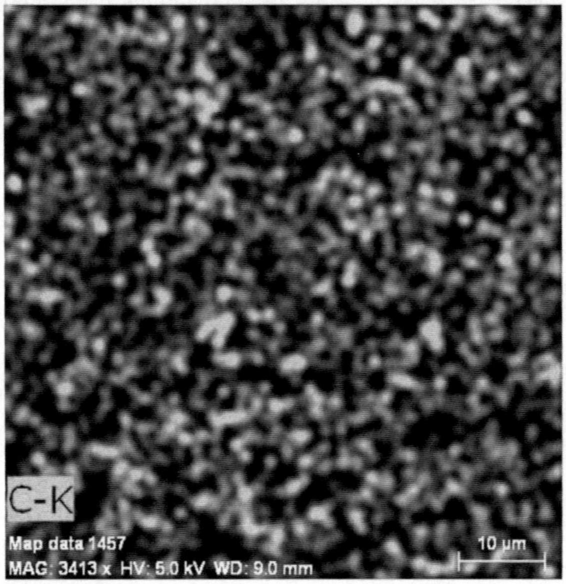

Figure 4. SEM image of deposited carbon on nickel substrate.

The comparison of spectra, registered in plasma jet in Ar/H_2 and $Ar/H_2/CH_4$ gas mixtures, confirms the effectiveness of the microwave discharge of the methane decomposition and the creation of numerous reactive species with high concentrations. The SEM image of carbon on Ni surface is presented in figure 4. It shows first step of growth of carbon nanostructures at substrate temperature of 650 C. Further investigations will include deposition at higher substrate temperatures and optimization of the deposition process.

4. Conclusion

Plasma reactor with microwave source works with argon, hydrogen and methane gas mixture at atmospheric pressure including an additional substrate heating. The registration of key reactive species (C, CH, Swan band) by optical spectroscopy and the achieved temperatures (650 C) prove the potential of the plasma reactor to produce specific carbon nanostructures.

Acknowledgements

This work is financially supported by Science Fund of Sofia University, project N114/2014. Authors thank Prof. Rusev for SEM image of substrate surface.

References

[1] Jiaxin Zheng, Lu Wang, Ruge Quhe, Qihang Liu, Hong Li, Dapeng Yu, Wai-Ning Mei, Junjię Shi, Zhengxiang Gao Jing Lu, *Scientific Reports* **3**, 1314 (2013)
[2] Wenjing Yuan and Gaoquan Shi, *J. Mater. Chem.* A,**1**, 10078 (2013)
[3] Cai M, Outlaw R, Butler S, Miller S, *Carbon*, **50**, 5481 (2012)

[4] Bo Zh, Yu K, Lu G, Wang P, Mao S, Chen J, *Carbon*, **49**, 1849 (2011)
[5] Tatarova E, Henriques J, Luhrs C, Dias A, Philips J, Abrashev A, *App. Phys. Lett.* **103**, 134101 (2013)
[6] Tatarova E, Dias A, Henriques J, Botelho do Rego AM, Ferraria AM, Abrashev MV, Luhrs CC, Phillips J, Dias FM and Ferreira CM, *J. Phys. D: Appl. Phys*. **47** 85501 (2014).
[7] Jasek O, Synek P, Zajickova L, Elias M, Kudrle V, *J Electrical Eng.* **61**, 311(2010)
[8] Kiss'ovski Zh, Djermanova N, Mitev D and Vachkov V. *J.of Physics: Conf. Series* **514**, 012007 (2014)

Influence of diaphragm configuration on DC diaphragm discharge breakdown in electrolyte solution

L Hlavatá[1], L Hlochová[1], Z Kozáková[1], F Krčma[1], O Guaitella[2]

[1]Brno University of Technology, Faculty of Chemistry, Purkyňova 118, 612 00 Brno, Czech Republic
[2]Laboratoire de Physique des Plasmas, Ecole Polytechnique, Route de Saclay,F-91128 Palaiseau, France

E-mail: xchlavatal@fch.vutbr.cz

Abstract. This paper deals with generation of diaphragm discharge in water solutions of sodium chloride with fixed conductivity of 275 μS. The photos were recorded by high-speed iCCD camera at defined times synchronized by the current passing through the system. Three different phases in the current-voltage curve were recognized under all conditions. Only electrolysis proceeds during the first phase at the lowest applied voltage (300 – 1200 V depending on the orifice dimensions), while bubbles were created thanks to the intense Joule heating inside the orifice during the second phase. Finally, the discharge was ignited at applied voltages over 1600 V. These facts were confirmed by iCCD images taken during all three phases. We concluded, by comparing current-voltage characteristics of different orifice diameter sizes, that this parameter had an important influence on the bubbles generation phase. On the contrary, the diaphragm thickness played an important role at the discharge breakdown.

1. Introduction

Pin hole discharge configuration consists of two electrode spaces divided by the dielectric barrier with a central pin-hole. Mechanisms of discharge breakdown in liquids are still under an intensive research and their study requires specific approach in all possible configurations (different kind of high voltage, various electrode configuration, etc.). Generally, two types of theories are considered to be the most suitable for discharge breakdown description – thermal (bubble) theory, and electron theory [1].

Discharge in the pin hole configuration probably combines both theories. Initially, it starts in the pin-hole when sufficient power is applied. The breakdown moment is probably related to the bubbles formation [2]. By the application of constant DC voltage, water solution is significantly heated due to high current density (Joule´s heating), and microbubbles of water vapor are created in the pin-hole region. It is assumed that the discharge breakdown starts inside these bubbles because of high potential gradient between the outer and inner bubble region [2], in correspondence to the thermal theory. On the other hand, further propagation of plasma channels to the bulk solution probably corresponds to the electron theory. Moreover, application of DC voltage initiates creation of two kinds of plasma streamers on both sides of the dielectric barrier [3][4].

2. Experimental setup

The batch plasma reactors using a diaphragm configuration with total volume of 100 ml are employed in the presented work. The discharge was created in an orifice (a pin-hole) in the dielectric barrier

Journal of Physics: Conference Series **768** (2016) 012018 doi:10.1088/1742-6596/768/1/012018

separating two electrode parts of the reactor. DC non-pulsed high voltage up to 2 kV was used for the discharge generation. Electrodes were made of stainless steel and they were installed in parallel to the diaphragm from the dielectric barrier in each reactor part. The dielectric barrier was Shapal-MTM – ceramics made with the variable thickness (0.3-2 mm). One pin hole at the diaphragm centre with diameter of 0.3-1 mm was used in experiments. Water solution containing NaCl electrolyt to provide initial conductivity of 275 $\mu S{\cdot}cm^{-1}$ was used as a liquid medium.

Time resolved characteristics of current, voltage and images camera were recorded using four-channel oscilloscope which detected their output values. Oscilloscope LeCroy LT374L (4-channels) operating up to 500 MHz with high voltage probe ELDITEST GE 3830 30kV, 1:1000 was used to obtain resolved characteristics of discharge voltage. Another probe 1:1 was used for current measurement using ballast resistor with resistance of 5.13 Ω. Simplified scheme of oscilloscope integration into the electrical circuit is demonstrated in figure 1. Mean values of breakdown parameters (voltage, current, power and resistance) were calculated and subsequently, static current-voltage characteristics were constructed for each experiment. We also used ICCD camera iStar, with program Andor iStar and objective Cosmicar/Pentax TV Lens 50 mm, 1:2,8, for the discharge imaging; camera synchronization was based on current passing through the system. We set either the positive or negative peak, which give signal over the Pulse Generator TTi, 10 MHz TGP 110 to the camera. The camera captures with records over 0.3 ms.

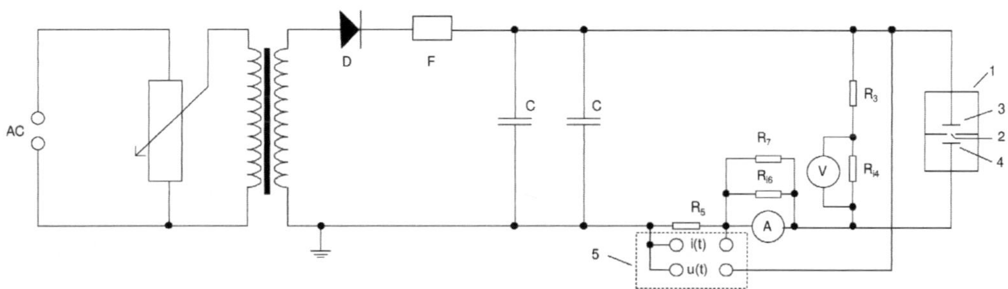

Figure 1. Simplified scheme of electrical circuit containing discharge reactor and oscilloscope: Electrical scheme of the experiment: 1 – discharge reactor, 2 – dielectric barrier with pin-hole, 3 – anode, 4 – cathode, 5 – oscilloscope LeCroy LT374L, F – fuse (0.5 A), C – microwave capacitor (0.85 μF), R – resistances: R_3 (100 MΩ), R_{i4} (3.114 kΩ), R_5 (5.13 Ω), R_{i6} (105.5 Ω), R_7 (0.13 Ω).

3. Results and discussion

Current-voltage characteristics of DC pin hole discharge were constructed from the mean values of time resolved current and voltage records over 10 ms. Figure 2 demonstrates a typical current-voltage curve obtained in NaCl solution with initial conductivity of 270 $\mu S{\cdot}cm^{-1}$. This curve could be divided into following three parts. Initially at low applied voltages, increasing the applied DC voltage, measured current increased slightly and more or less directly proportionally. The time resolved current record depicted on figure 3 shows relatively smooth line and thus we can conclude that only electrolysis took place in the system. Of course, the electrolyte solution was heated by the Joules effect due to the passing current. Going above the voltage of some hundreds volts, the first significant breakpoint appeared in the curve – current gently jumped up. According to the time resolved characteristics and photo demonstrated in the figure 4(a), the smooth current time record changed and some current pulses can be recognized. These current pulses were related to the substantial creation of micro bubbles formed by the evaporating solution inside the pin hole where the current density was the highest and thus the Joule heating was sufficient for microbubble creation as shown in the photo. Further enhancement of voltage provided only a small current increase until the second remarkable breakpoint was observed. From this moment, current was rapidly arising with only a small voltage enhancement. This second breakpoint was assigned to the discharge breakdown moment and with further voltage increase discharge started to operate more regularly.

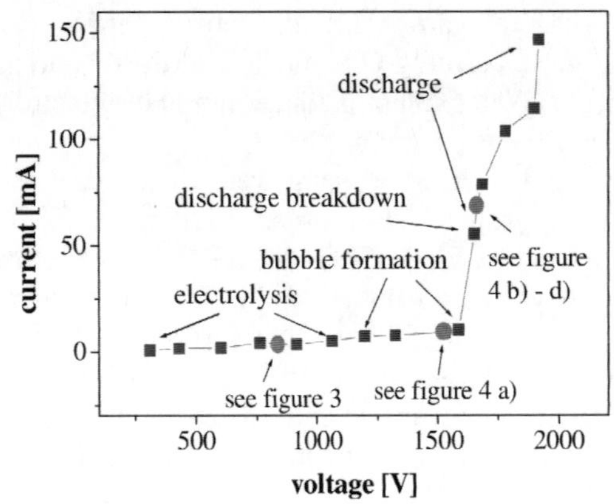

Figure 2. Typical current-voltage characteristic of DC diaphragm discharge in NaCl solution with initial conductivity of 275μS·cm^{-1} and 0.7 mm barrier thickness with pin-hole diameter of 0.3 mm.

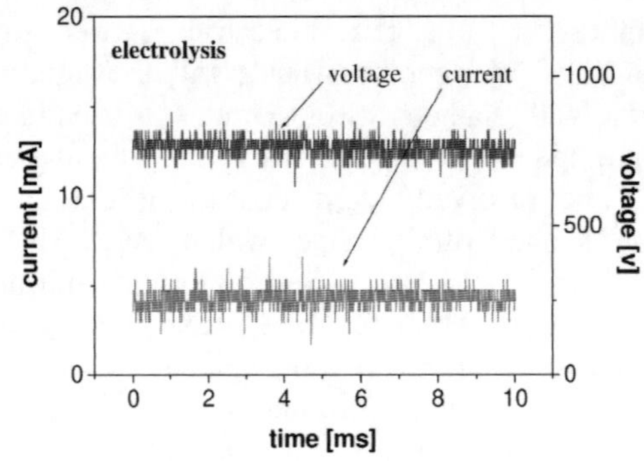

Figure 3. Time resolved current record in NaCl solution with mean voltage values of 860 V.

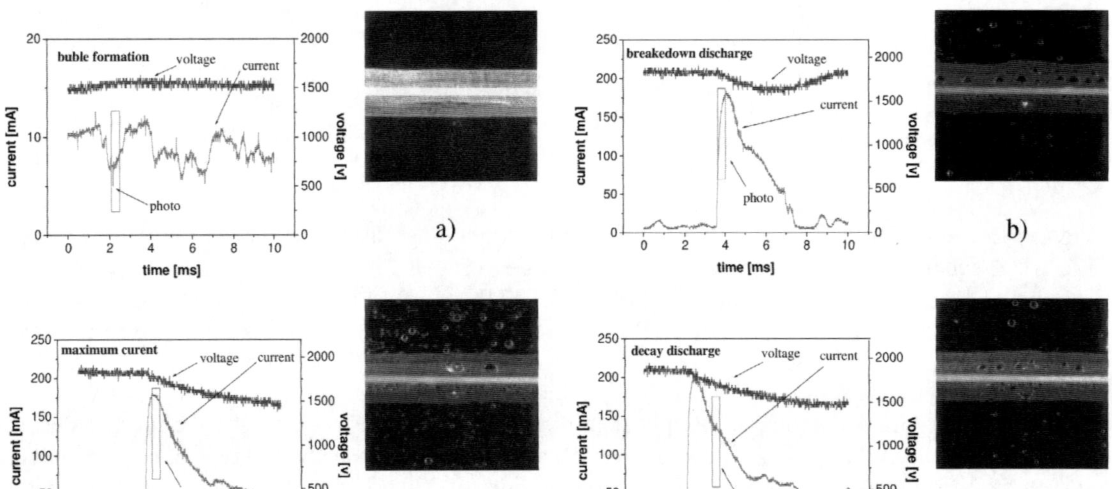

Figure 4. Time resolved characteristic of voltage and current with photo in NaCl solution: mean voltage values of (a) 1530 V; (b) – (d) 1700 V

Time resolved characteristics and photos for different discharge phases are demonstrate in figures 4 (b) – 4 (d). Figure 4 (b) demonstrates random discharge breakdown with light emission at the current peak maximum as it is shown on the photo. Figure 4 c) demonstrates bubbles with ignited discharge created around the pin-hole when current reached its maximum. The photo in figure 4 (d) demonstrates that current decrease is followed by weakening of light emission and subsequent bubble expansion.

As it is shown in figure 5 in phase of electrolysis the current through the pin-hole is higher with the pin-hole diameter increasing at constant voltage and diaphragm thickness. Start bubbles formation were slightly increased from 700 V for diameter of 0.3 mm to 900 V for diameter 0.6 mm and 1000 V for diameter 1 mm. Breakdown voltage as well as current were increased with increasing the pin-hole diameter. The measured breakdown voltage for 0.3 mm pin-hole diameter is 1200 V; for diameter 0.6 mm is 1300 V; for diameter 1.0 mm is 1400 V.

The increase of the barrier thickness has a substantial effect only on the discharge breakdown part of the current-voltage curve as it is shown in figure 6. During electrolysis and bubble formation phase the increasing of the barrier thickness has a negligible effect on the current. However, the breakdown voltage increases from 1300 V (at thickness of 0.3 mm) to about 1500 V for 0.7 mm thickness and to about 1800 V for 1.5 mm thickness.

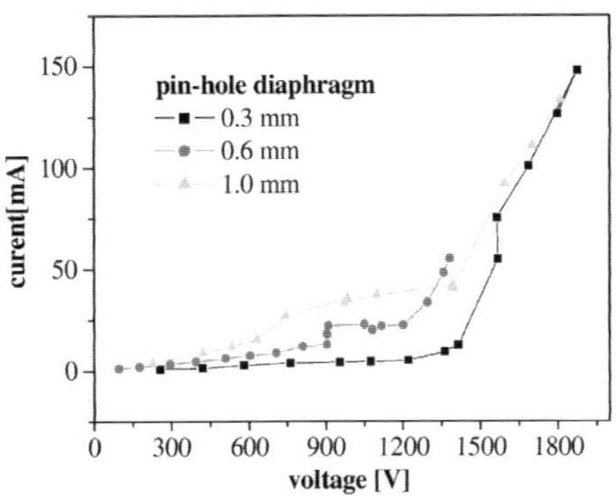

Figure 5. Comparison of current-voltage characteristics of the diaphragm discharge in NaCl solutions with conductivity of 275 μS·cm^{-1} for three pin-hole diameters diaphragm.

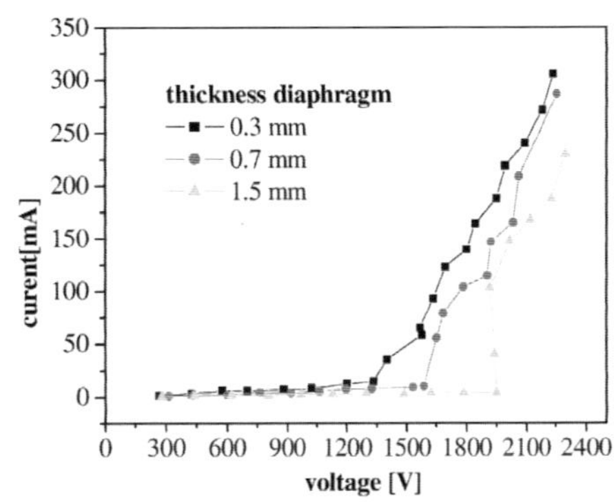

Figure 6. Comparison of current-voltage characteristics of the diaphragm discharge in NaCl solutions with conductivity of 275 μS·cm^{-1} for three different barrier thicknesses.

4. Conclusion

Breakdown moment, as well as processes of electrolysis and bubbles formation were identified from obtained statistic and time resolved characteristics of DC diaphragm discharge and time resolved photos. Current-voltage evaluation was remarkably influenced by using different pin-hole diameter and barrier thickness. Increase of diaphragm diameter has a major effect on the electrolysis. Increase of diaphragm thickness determines discharge breakdown as well as its operation.

Acknowledgements
Presenting author thanks to the ERASMUS student exchange scholarship.

References

[1] Malik M A, Ghaffar A, Malik S A 2001 *Plasma Sources Sci. Technol.* **10** 82.
[2] Joshi R P, Qian J, Schienbach K H 2002 *J. Appl. Phys.* **92** 6245
[3] Stará Z, Krcma F, Procházková J 2008 *Acta Technica CSAV* **53** 277.
[4] Hlavatá L, Krcma F, Kozáková Z 2011 *Book of Contributed Papers: 18th Symposium on Application of Plasma Processes and Workshop on Plasmas as a Planetary Atmospheres Mimics*, Vrátna 2011, pp. 83–87 (id 18960)

Increase atom/molecular ratio of the hydrogen discharge

D V Kolodko, N V Mamedov, I V Vizgalov, D N Sinelnikov, I A Sorokin

Plasma Physics Department, National Research Nuclear University MEPhI, 31 Kashirskoe shosse, Moscow, 115409, Russia

E-mail: dobrynya_kol@mail.ru

Abstract. The high atomic to molecular ion ratio is important for hydrogen ion sources, but in the usual low-temperature hydrogen plasma the molecular ion component is dominating. Proton component can be increased by means of increasing plasma temperature, but in this case energy spectrum will be widened. Single electron collision with a hydrogen molecule may not only lead to the formation of the molecular ion, but also results in the formation of the proton. However, the cross section of the second process is much smaller than of the first (approximate ratio between them being 1:200). Thus, a single electron-molecule collision cannot be an efficient means of proton formation. Atomic hydrogen ions are produced by ionization of neutral hydrogen atoms. As a rule, a lifetime of neutral atoms is too short for ionization. The increase of the hydrogen atom lifetime by reducing the wall recombination coefficient allows us to get more protons. In this work we study the influence of water vapor injection on atom/molecular ratio of hydrogen plasma.

1. Introduction

Nowadays ions sources have a large range of applications. As a rule, ion sources consist of a plasma source and an electro-optical system. Ions are drawn from plasma by accelerating potential and ion beam is formed by optic system. If plasma is made of molecular gas (like hydrogen) the ion beam will contain atom and molecular components. The cold plasma created by compact Penning source usually has a low atom/molecular ratio [1]. Atom component can be increased by means of increasing plasma temperature, but in this case energy spectrum of the ion beam will be widened. And in the compact sources it is impossible to increase discharge power.

In particular applications, for example, in Gas-filled Neutron Tube (GNT), usually only the atom component is needed. Energy spectrum of ions from Penning Plasma Source (PPS) (which is used in GNT) is a narrow peak and all particles have the same energy. So, the energy of deutron is twice higher than the energy of individual deutron bound in molecule ions. The increase in atom component from 5% to 35% can make GNT 3 times more effective, and 100% of atom component increases GNT efficiency sevenfold.

2. Main elementary processes of ions formation in a hydrogen plasma

Single electron collision with hydrogen molecules can lead to the generation of molecular ions or protons [2]:

$$H_2 + e = H_2^+ + 2e - 15.4\ eV$$

$$H_2 + e = H^+ + H + 2e - 18\ eV$$

But interaction cross-section of second process is far less than that of the first (their ratio being 1:200) [3]. Thus, a single electron-molecule collision cannot be effective for generation of protons. In the low temperature plasma the main processes of atomic hydrogen ions formation are [2]:

1. Dissociation of molecules with subsequent ionization of atoms:

$$H_2 + e = H + H + e - 4.7\ eV;$$

$$H + e = H^+ + 2e - 13.6\ eV;$$

2. Ionization of molecules with subsequent dissociation of the molecular ion:

$$H_2 + e = H_2^+ + 2e - 15.4\ eV;$$

$$H_2^+ + e = H^+ + H + e - 2.6\ eV;$$

$$H_2^+ + e = H^+ + H^+ + 2e - 16.2\ eV;$$

3. Dissociative recombination:

$$H_2 + e = H_2^+ + 2e - 15.4\ eV;$$

$$H_2^+ + e = H + H + 11.0\ eV;$$

$$H + e = H^+ + 2e - 13.6\ eV;$$

Thus, formation of protons in plasma normally goes through ionization of hydrogen atoms. But lifetime of hydrogen atoms is very short because there is no confinement by magnetic field and they recombine on the walls. Thus, their lifetime strongly depends on recombination coefficient of plasma-facing components.

Table 1. Recombination coefficients for various materials.

Material	r_H (recombination coefficients)
Silex	7×10^{-4}
Pyrex	$7{,}5\times10^{-4}$
Aluminium (Al_2O_3)	10^{-3}
Copper	0,19
Iron	0,17
Nickel	0,18

Table 1 shows recombination coefficients for various materials. Materials with a lower coefficient are usually dielectric. Thus, if plasma-facing components are coated with a dielectric layer, then lifetime of hydrogen atoms will increase.

3. Experimental study of the influence of water vapour injection on atom/molecular ratio

The easiest way to modify plasma-facing components and decrease recombination coefficients is to form a dielectric oxide layer.

3.1. Experiment in mirror machine PR-2

The experiment on PR-2 [4] shows that small water vapor injection can significantly change ion spectrum. In this case intensity of impurities peaks is less than 5% and the partial pressure of water vapor is less than 3% of total pressure.

The increase in atom/molecular ratio takes place only after 30 minutes after the thermo-chemical water vapor source (TCWVS) [5] is turned on and remains after shutdown of TCWVS for a long time. This delay points onto the wall surface processes decreasing the recombination coefficient.

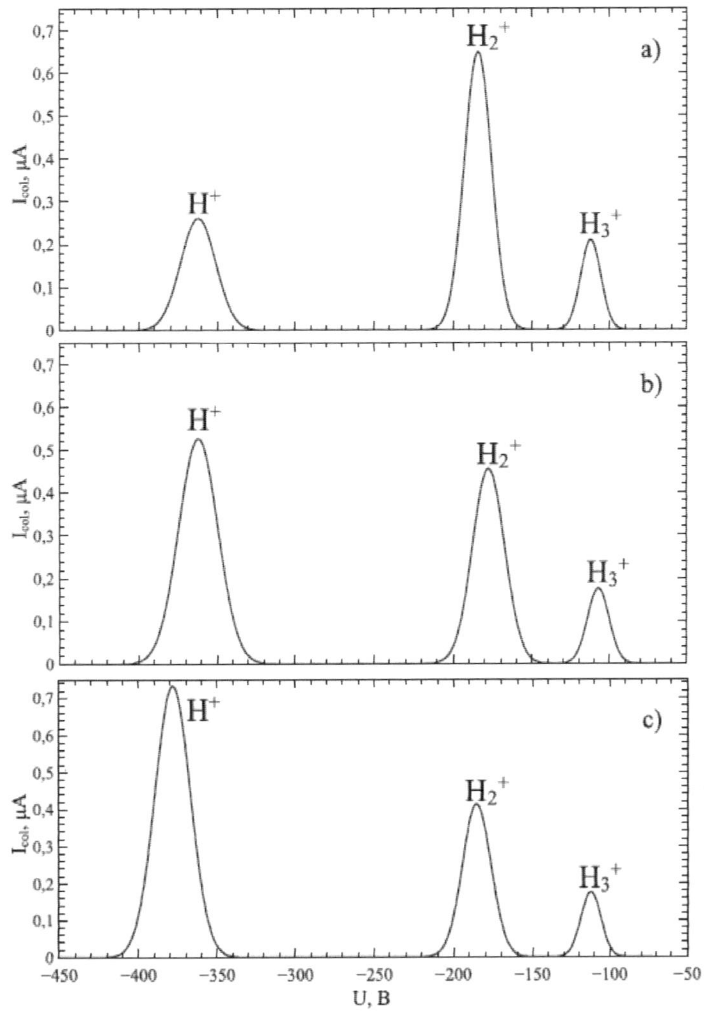

Figure 1. Ionic composition of hydrogen plasma in the PR-2:
a) without water vapor injection;
b) after 30 minutes, TCWVS is on;
c) after 1 hour, TCWVS is off

3.2. Experiment in the installation for studying small-sized ion sources:
The same experiment was performed on a small-sized PPS [6]. Plasma-facing components of PPS are made from stainless steel which can form the oxide layer, but our experiments have shown that this layer is nonsufficient to achieve the expected effect.

4. Conclusion
The materials of plasma-facing components with a lower recombination coefficient (dielectrics) can lead to increase in atom/molecular ions ratio in hydrogen plasma. The simplest and the cheapest way

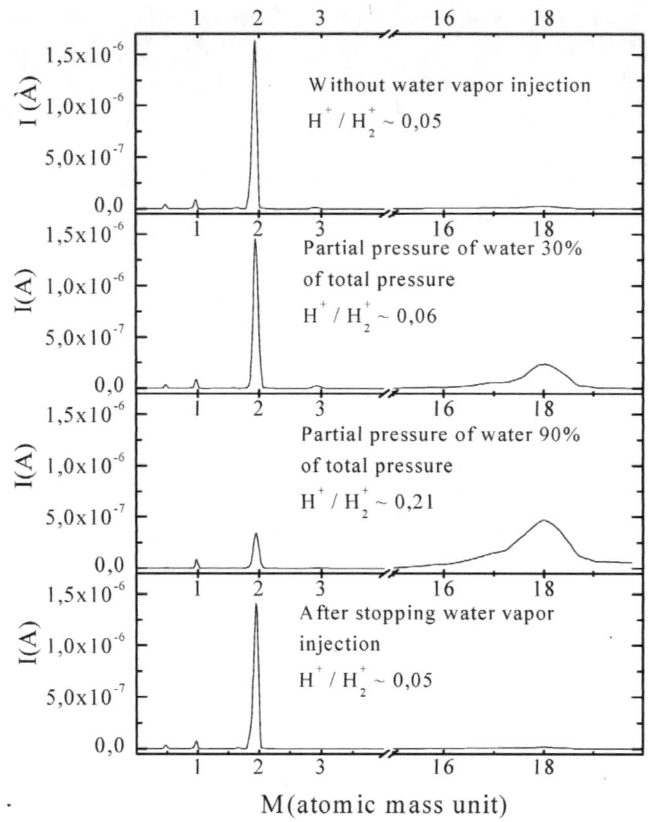

Figure 2. Relations of mass-spectum and water vapor injecting.

to change recombination coefficient is injection of small admixtures of oxygen or water vapor into a chamber of hydrogen plasma source. For this it is necessary to use materials which form dielectric oxide layers with low recombination coefficients. Thus the lifetime of hydrogen atoms and, correspondingly, the atomic ions output from the source can be increased.

References

[1] Forrester A T 1989 *Large ion beams, fundamentals of generation and propagation* (New York: Wiley)
[2] Gabovich M D 1971 *Plasma physics and technology of ion sources* (Moscow: Atomizdat)
[3] Newhall H F 1942 *Physical Review* C **62** 11
[4] Visgalov I V, Gutorov K M, Pisarev A A 2007 A mechanism of PFM erosion and redeposition in gaps *Journal of Nuclear Materials* **363-365** 966-971
[5] Visgalov I V, Gutorov K M, Podolyako F S, Sorokin I A 2011 Development and testing of the thermochemical source of water vapor to simulate the cooling system leak *Plasma Physics department, annual report* (Moscow: MEPhI)
[6] Mamedov N V, Kurnaev V A, Visgalov I V, Sinelnikov D N 2013 *NUCLEAR PHYSICS AND ENGINEERING* **4** (Moscow: Nauka / Interperiodica) 436–442

Institute of Physics
Dirac House, Temple Back
Bristol BS1 6BE UK

ISSN: 1742-6588
ISBN 978-1-5108-3227-5
9 781510 832275

8th National Scientific Conference on Advances in Electrotechnology (Postepy w Elektrotechnologii)

IOP Conference Series: Materials Science and Engineering
Volume 113

Jamrozowa Polana, Poland
23 – 25 September 2015

IOP
Institute of Physics